Facts not Opinions

For Denis Smith, who brought
Kirkaldy's name to life once more.

150 years of testing which changed the world...

FACTS NOT OPINIONS

By Tim Crichton, Lucy Hogarth, Colin Jenkins and Hugh MacGillivray

Foreword by Marty Jopson

Foreword by Marty Jopson

Above: Marty Jopson.

Since 2007 it has been my very great pleasure to be the science and engineering reporter for the BBC1's One Show. During that time I have travelled the length and breadth of the country meeting lovely people, discovering fascinating science and filming in extraordinary places. Few of these places, though, are as enchanting or unexpected as the Kirkaldy Testing Museum, situated as it is just south of the river in the heart of our capital city.

In 2016 we filmed at the museum for the One Show, but this was not my first encounter with David Kirkaldy. Previously I had spent a decade making science television from behind the camera and Kirkaldy's story had featured in another series I was involved in making. That time, however, I did not get a chance to see the Universal Testing Machine in action so I was delighted to be filming at 99 Southwark Street once more, and to have the chance to bring this special story to the One Show audience. Although, I must admit, part of that delight was knowing that this time I would get to witness some testing in action.

My experience filming at the museum this time convinced me that it is a magical place, nestled as it is amongst the towering glitter of bustling Southwark. Suddenly, you find yourself in an oasis of calm, a delicious counterpoint to all that is modern outside.. As we filmed with the doors wide open I watched passing visitors peering in surprise at the splendid Victorian machinery within.

This is the real joy of the museum. Kirkaldy's machine, so lovingly cared for by the museum volunteers is crucially more than just a museum piece: it still works, and to see it in action is truly special. While filming the One Show we used the machine to put some wrought iron bars to the test and as I watched the skilful operation by volunteers Roz and Lucy, the building tension was palpable, both literally and metaphorically! I realised then just how lucky we are that the museum exists and continues to keep alive the work and legacy of David Kirkaldy.

I realise this sentiment is just my own opinion, though one I think that few would argue with. However, the historical significance and importance of the work started by David Kirkaldy and carried on by his son and grandson at their Testing and Experimenting Works is, without doubt, a fact.

How the Kirkaldys Changed the World

Above: the Southwark-Emery Universal Testing Machine build in 1929. *Photograph by Ab1247, 2016, Creative Commons Attribution-Share Alike 4.0 International license.*

What is Materials Testing and Why is it Important?

Materials Testing is an area of engineering science that quantifies the properties of materials used in construction (e.g. their strength under tension, compression etc...). This can range from the beams used in a bridge to the propellers used in an aircraft engine. Despite being a relatively young science, Materials Science is one of the most important areas of research in the scientific world. Comparative research and tests have allowed the creation of rigorous standards and new forms of testing for all building materials. Without these standards, there would be no criteria of quality for construction materials to which constructors have to adhere.

During the Industrial Revolution in the 19th century, the need for these standards was beginning to become apparent. Designs of large bridges, ships with powerful engines and extensive railways were being conceived to meet the need for national infrastructure, which would potentially deliver thousands of people a day. The result of the creation of these major projects was sometimes disastrous. The collapse of the Tay Bridge is a famous and demonstrative example of how the lack of materials standards can lead to tragedy. The investigation into the demise of the Tay Bridge (an inquiry in which David Kirkaldy was involved) found that many of the cast iron columns had holes within them, which had been crudely filled with some sort of putty. As there were no sort of criteria for those who cast the iron columns, they had performed the casting poorly and, presumably, at the lowest expense to themselves.

It was incidents such as the fall of Tay Bridge and the foresight of engineers like David Kirkaldy that led to the adoption of Materials Science as a integral part of any construction process.

How did David Kirkaldy Change Materials Testing?

In 1862, David Kirkaldy published his treatise on the testing of iron and steel entitled "Results of an Experimental Inquiry into the Tensile Strength and other Properties of various kinds of Wrought Iron and Steel". This work was the culmination of his scrupulous record making and detailed experiments that he had undertaken up to that date. Furthering these investigations, Kirkaldy designed his Universal Testing Machine. This was the first testing machine that was universal; it was capable of performing a multitude of different tests on a single specimen. His machine has been a protype for many of its successors over the years.

Kirkaldy's unwavering determination to have testing standards enforced was initially met with disdain, but gradually his work was given the credence it deserved. David Kirkaldy's "Testing and Experimenting Works" where he performed commercial testing and his own investigations gained international renown within Kirkaldy's lifetime.

The son and grandson of David Kirkaldy, William George and David respectively, continued to operate the testing works and upheld the association of the Kirkaldy name with precision and fortitude. Within this book, the stories across the world that made the name of David Kirkaldy's testing works will be told. These stories changed the face of construction engineering forever and has impacted every construction project you will see today.

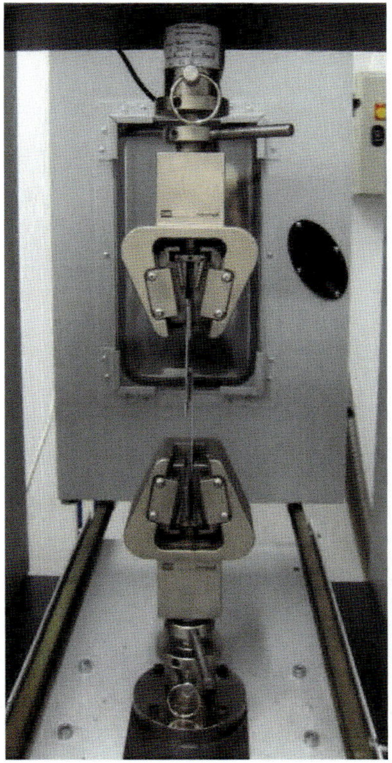

Above: modern tensometer equipped with wedge action jaws.

A Chronology of the Works

1863 David Kirkaldy patents the design for his Universal Testing Machine.

1865 The machine is built by Greenwood & Batley of Leeds and sent to London. Kirkaldy then starts work in The Grove, Southwark.

1866 The Works is officially opened on the 1st January in The Grove.

1867 Kirkaldy tests parts for James Eads 1867 St Louis Bridge over the Mississippi River.

1874 The Works is built and opened in its current location at 99 Southwark Street.

1880 Kirkaldy is asked to perform tests on parts of the failed Tay Bridge.

1887 The chains and bars for the Hammersmith Bridge are send to the Works.

1897 David Kirkaldy dies and the Works is taken over by his son, William George Kirkaldy.

1914 William George's widow, Annie, runs the Works with Dr Gulliver as manager.

1932 Parts of the steel framework for the Sydney Harbour Bridge are sent to the Works.

1934 David Kirkaldy, William George's son, joins the Works and takes over as manager

1951 Parts for the Skylon for the Festival of Britain are tested at the Works.

1954 Parts of the de Havilland Comet are sent to the Works for testing after it crashes.

1965 David Kirkaldy retires and the Works continues under the management of Treharne and Davies.

1974 The Greater London Industrial Archaeology Society visits the Works and later the building and machine are listed for preservation.

1978 – 1983 The first museum proposal is made, and the Kirkaldy Testing Museum is set up as a charity.

Previous page: photograph, Museum of Fractures, 2nd Floor, Kirkaldy's Testing and Experimenting Works, 1885.

The Kirkaldys

Above: an engraving of David Kirkaldy

"He gave himself up entirely to engineering".

The Inst. C.E. Vol. CXXVIII, 1897

David Kirkaldy 1820 – 1897

> *"Cautious to a degree; enthusiastic past belief; honest as the sun; outspoken and fearless as a Viking, he laboured in a field previously untilled . . . Kirkaldy was the terror of so many persons, that it is probable that at one time he was the best hated man in London."*
>
> The Engineer, 1897

David Kirkaldy was born on the 4th April 1820 near Dundee in Mayfield. He was the second son of William Kirkaldy (1785-1858), who was a merchant and shipper within the city. David is described in his obituary, published by the Institute of Civil Engineers (ICE), as a boy with a "nervous and highly sensitive disposition which ... remained a characteristic throughout his life". He was educated at Merchiston Castle, Edinburgh, where he discovered a "taste for experimenting" and dedicated a Saturday afternoon to experimental work. After a time working under his father as a merchant, he began to find the occupation unsatisfactory, despite his recorded efforts to please his father. In 1843, he began an apprenticeship at Robert Napier's Vulcan Foundry Works at Lancefield in Glasgow, thanks to the support of a Mr. Erskine of Linlathen. His skill as a draughtsman led to his being moved, in 1847, to the drawing office, where he was swiftly promoted to "Chief Draughtsman and Calculator".

Early on in his career Kirkaldy would meticulously collect and note data on all vessels being built at Napier. In addition to this, he developed a finesse for engineering drawings, not only as a diagrammatic form but also as artwork. Throughout his time at Napier,

Kirkaldy produced a wealth of beautiful drawings of ships and machinery, using the records in his notebooks. In 1850, he completed coloured sketches of four liners, which were subsequently displayed at the Paris Exhibition of 1855 and later placed in The Louvre. Further to this, in 1861 Kirkaldy's drawing of RMS Persia was exhibited in the Royal Academy during its Summer Exhibition, the only time this honour had been bestowed upon a diagrammatic sketch of this kind. In this drawing, Kirkaldy said he had attempted to "combine artistic effect with the strictest accuracy in all the mechanical details". In both instances, he was awarded a medal for excellence and in the case of the liner sketches, Napier received a gold medal and Legion of Honour as exhibitor.

> *"Having a strong artistic bent, he took up drawing with keenness and, striking out a new line, astonishing engineers with his productions of highly finished and artistically coloured drawings".*
>
> *The Inst. C.E. Vol. CXXVIII, 1897*

Kirkaldy's meticulous attention to detail and systematic recording of the details of construction and performance of every ship and engine at Napier's, as well as his "earnestness and enthusiasm in this pursuit", brought him into trouble with his colleagues. His obituary written by ICE suggests that he was subject to prejudice and resentment as a result of his keenness for introducing his own improvements. This bitterness is recorded to have got to a point where Napier was urged to dispense with many of David Kirkaldy's projects. Most notably this included his "trial-trip" initiative, which aimed to hold shipbuilders to more rigorous standards. Despite this, Kirkaldy continued to accumulate vast amounts of data during his time at Napier.

In 1858, on the 2nd June, Kirkaldy married Annamelia Yates Miller, by whom he had a son and daughter. By this time, he had become a well established employee of Napier and had aided the establishment of the Institution of Engineers in Scotland.

Kirkaldy's interest in materials testing seemed to flourish under the Napier tests that occurred in the period 1858 to 1861. The firm had received orders to constructing high-pressure boilers and marine machinery where the choice of materials would be crucial.

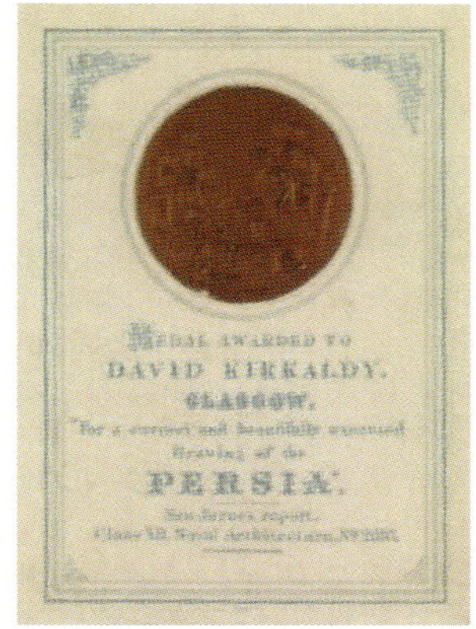

Above: medal awarded to David Kirkaldy, Glasgow. "For a correct and beautifully executed drawing of the Persia".
© National Maritime Museum.

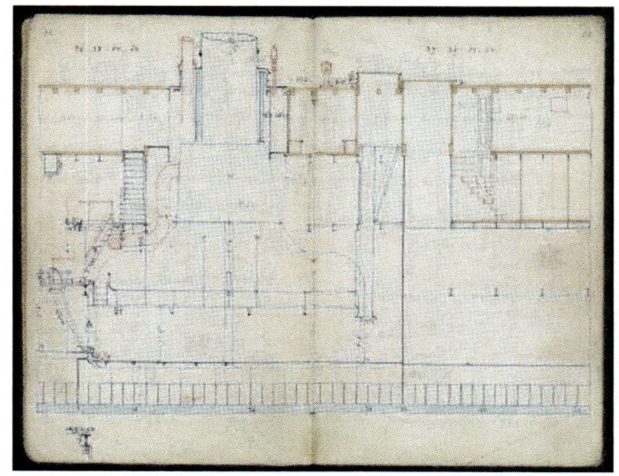

Above: notebook containing preparation drawings of the construction and fitting for 'Persia' (1855).
© National Maritime Museum.

RMS Persia

In 1852, Samuel Cunard, founder of the Cunard Line across the Atlantic, ordered a new liner, the Arabia, to be constructed in order retake the Atlantic crossing records from his competitors. However, due to being a relatively small ship with a wooden hull, the Arabia was unable to cope with the powerful engines with which it was installed. In response, Cunard elected to construct a ship with an iron hull.

Napier and Sons of Glasgow was commissioned to design a ship with an iron hull to replace the Arabia in the Cunard fleet. However, the ship they conceived was 50% larger than the Arabia making her the largest ship in the world at the time. This ship was named the Persia and she was the first Atlantic record breaker made of iron. She took the Atlantic speed record in both directions, her best time being 9 days, 10 hours and 22 minutes when heading eastwards.

The Persia's two-cylinder side-lever engine produced 3,600 horsepower and, at her launch in July 1855, she reached 17 knots (31 km/h). However, due to her high rate of fuel consumption, 147 tons per day, her speed was limited to 13 knots (24 km/h). It was this high demand for fuel that was the eventual downfall of the Persia. In 1863, Cunard commissioned the Scotia, which took the Atlantic record from the Persia. As screw propulsion was side lined, and further vessels added to Cunard's fleet, the Persia was decommissioned due to her fuel consumption. Attempts to convert her to sail failed and she was sold for scrap in 1872.

Left: British & North American Royal Mail steam ship RMS Persia ,1855.
© National Maritime Museum.

Below: engraving of RMS Persia.
Unknown, date unknown.

Far below: side-lever engines of the RMS Persia and RMS Arabia, side elevation.
Unknown, c.1860.

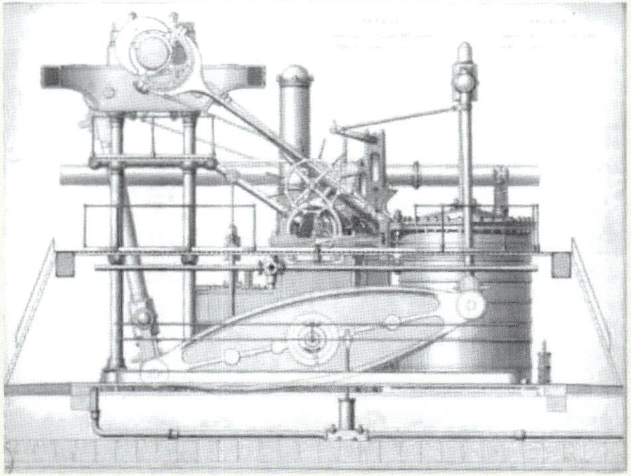

The page and opposite: RMS Persia under construction at the Napier and Sons shipyards in Glasgow.
Berwick & Annan, DeGolyer Library, Southern Methodist University, 1855.

THE PERSIA,

Built by R. Napier & Sons, 5260 Tons displacement, Launched 3d July 1855.

HMS Dauntless

The Dauntless was a steam screw frigate, which launched on 5th January 1847. It was the creation of John Fincham who initially intended her to be a paddle vessel. She was only ever partially redesigned to accommodate the screw propulsion, which adversely affected her performance. Under steam, the Dauntless had a best speed of 10.3 knots (19.1 km/h) and, despite being lengthened in 1850 to better suit her original framework, her performance remained unchanged. The disharmony between the paddle and screw driven drives continued to plague the Dauntless throughout her service.

The career of the Dauntless saw her assigned to a range of fleets across the world. Her first was with the Experimental Squadron in 1850, which employed vessels of unorthodox design. Later, in 1852, she was seconded by the North America and West Indies Station.

At the start of the Crimean War in 1854, the Dauntless served in the Baltic and later in the Black Sea, at Eupatoria, at the bombardment of Sevastopol and finally at Kinburn. In 1859, she retired to become the Coastguard base ship at Southampton. The Dauntless was sold for scrap in 1885 after being laid up at Devonport in 1878.

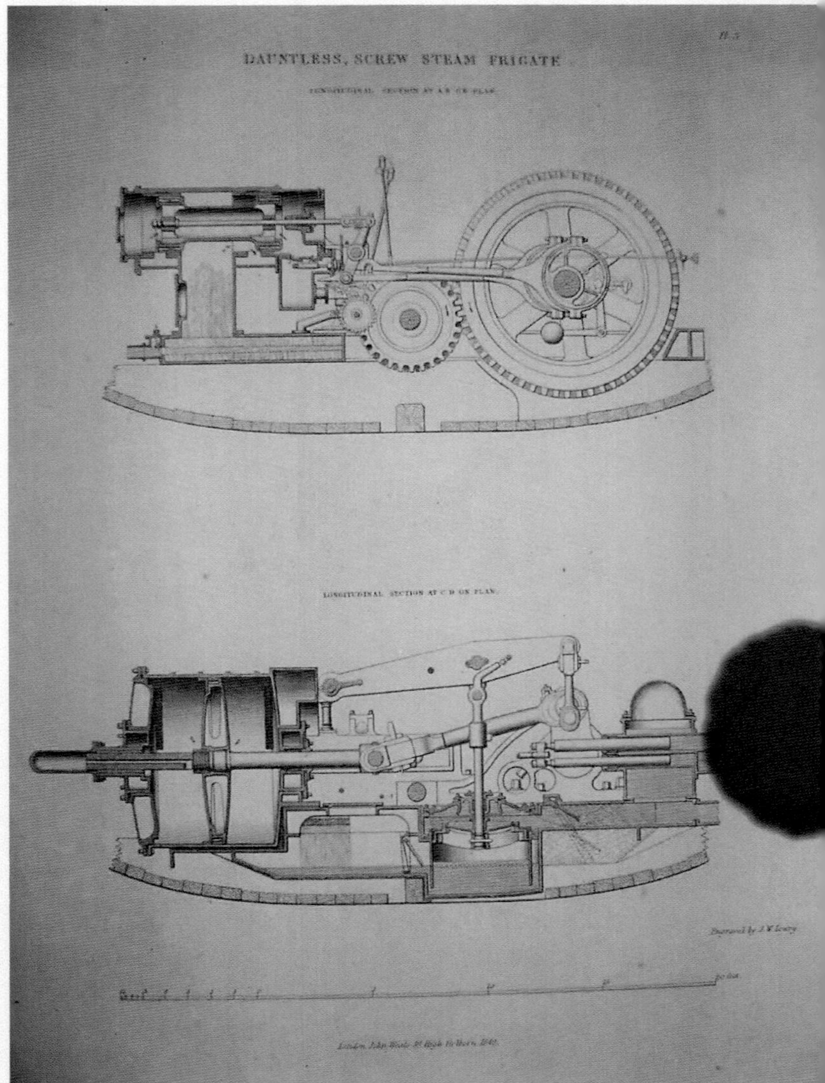

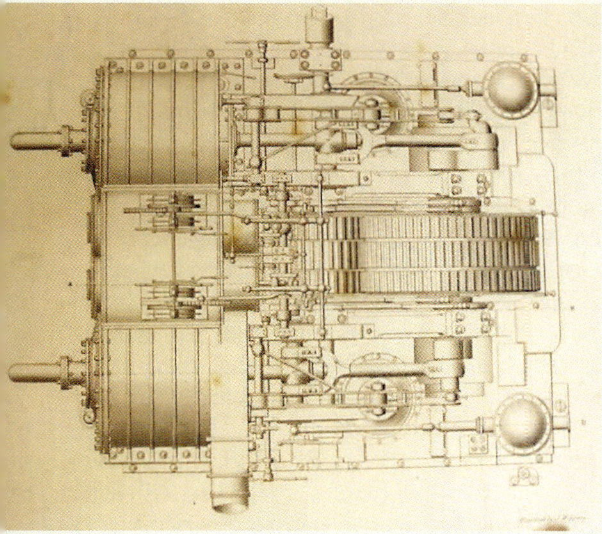

Opposite left: lithograph of HMS Dauntless, published by Day & Son, Lithographers to The Queen, c. 1855.
Lithograph by T G Dutton of Day & Son, c. 1855.

Opposite right: sketch of "HMS Dauntless Screw Steam Frigate" by David Kirkaldy.
© Science Museum / Science & Society Picture Library.

Above: sketch of "Engines of HMS Dauntless Screw Steam Frigate" by David Kirkaldy.
© Science Museum / Science & Society Picture Library.

Below left: engraving of Robert Napier.
Engraving by William G. Jackman, date unknown.

Below right: past presidents of the Institution of Mechanical Engineers in 1866. From left to right: Penn, Whitworth, Napier and Fairbairn.
Unknown, Manchester, 1866.

Robert Napier

David Kirkaldy received much of the inspiration for his work at Napier's Vulcan Foundry. He was among many of Scotland's most eminent shipbuilders who were also apprenticed under Robert Napier (the founder of Napier). Among these greats were James and George Thomson of the J&G Thomson shipyard and John Elder of the Fairfield Shipbuilding and Engineering Company.

Robert Napier founded his Vulcan Foundry in Glasgow in 1828. For the most part, his firm built steamship engines and steam engines for seafaring vessels. Napier's engines were both more reliable and more economical than his competitors on the Thames, which saw him appointed as the Admiralty's primary engine builder in 1838. Later, Napier would co-find the British and North American Royal Mail Steam Packet Company with Samuel Cunard. The RMS Persia, launched in 1855, was the largest ship in the world and was built to sail this line.

He received further prestige in 1849, when he built the Leviathan, which was the first train ferry between Granton and Burntisland. Seven years later he built HMS Erebus, the Royal Navy's first iron-clad vessel.

Robert Napier was appointed a member of the Institution of Civil Engineers and later became President of the Institution of Mechanical Engineers.

"He was so scrupulous, even jealous, in maintaining the line of action which he had marked out for himself that be could not easily brook dictation or thwarting".

The Inst. C.E. Vol. CXXVIII, 1897

Above: HMS Black Prince
Unknown, US Navy Historical Centre, 1861.

David Kirkaldy was chosen to devise apparatus and conduct experiments for this project, predominantly to make a comparison between recently introduced homogeneous metals and puddled steel, instead of the usual wrought iron. He is recorded to have taken on the work with characteristic eagerness and became so interested by the project that he requested it be extended to a larger and more detailed study, for which he designed and erected his own testing machine. Using this equipment, Kirkaldy also tested the iron plates and angle-bars used in the armour-cased ships the H.M.S. Black Prince and H.M.S. Hector, work which resulted in a scientific paper. However, he was becoming disillusioned with some of the practices at Napier. He commented during the project that "all the pieces were taken promiscuously from Engineers' or Merchants' Stores" and also felt that measurements were imprecise, using fairly rudimentary methods. In September 1861 Kirkaldy completed the materials tests required for the project but realising Napier would not support the continuation of his research, he resigned from the firm.

Following his departure, in 1862 Kirkaldy published the book which made his name in the engineering world, based on the work that he had conducted at Napier. His treatise entitled "Results of an Experimental Inquiry into the Tensile Strength and other Properties of various kinds of Wrought Iron and Steel" was held over the next few decades as "the most comprehensive and thorough ever made" across the globe. Following his research, Kirkaldy also took out a provisional patent on the process of oil-hardening on the strength of steel, a process he had thoroughly investigated. Furthermore, in response to his work, he was awarded the Gold Medal of the Institution of Engineers in Scotland, giving him a firm position from which to launch his independent testing career. His study into the "effect of shapes" on the behaviour of metals in his treatise also led to the general adoption of a new type of bolt for securing armour plating on battle ships.

Over the next two and a half years, Kirkaldy began work on a Universal Testing Machine (to be constructed at his own cost) that would allow him to further his investigations in more depth. The functions are best summarised in Kirkaldy's own words:

"I designed my Patent Testing-Machine upon a large scale expressly for the purpose of testing all kinds of constructive materials, and that under the various stresses, namely: Pulling, Thrusting, Bending, Twisting, Shearing, Punching, and Bulging."

On 8th June 1864, Kirkaldy visited Messrs. Greenwood and Batley in Leeds in order to discuss the machine's construction, the actual order being made on the 27th. Following this order, there were many disputes concerning both the price and rights over the machine leading to Kirkaldy, after fifteen months, issuing the firm with six days to deliver the machine, completed or not. This delay in construction caused Kirkaldy "great anxiety" as well as significant monetary expense. The machine was hastily despatched in September 1865 and Kirkaldy founded his first works in The Grove, Southwark. Kirkaldy opened his Works for business on 1st January 1866.

Kirkaldy's testing works faced considerable opposition and prejudice in its formative years from those opposed to his new methods. However, still being foremost an experimentalist, Kirkaldy used this time to carry out research using his Universal Testing Machine. His work gradually led to important modifications being widely adopted in bridge-work, welding and railway construction, among many other areas. He received particular opposition to his proof that "punching" holes in steel as opposed to drilling, as was common practice, had a severely detrimental effect on the integrity of the metal.

As Kirkaldy's methods became more widely accepted, he decided to acquire larger premises for his testing works. In November 1872, the Metropolitan Board of Works agreed to "the erection of one building…and to the erection of a Steam Engine and Two Furnaces for use from Mr. Kircaldy (sic) in his business of testing building materials". Once granted this building, Kirkaldy employed T. Roger Smith, of 23 Bedford Street, Covent Garden, as his architect. No expense was spared in this project as Kirkaldy strived to create not only a business environment, but several experimental departments where he could continue his work as an ever enthusiastic experimentalist. A contract

"He fearlessly and consistently acted up to what appeared to him to be right, and he lived long enough to see his work widely acknowledged ".

The Inst. C.E. Vol. CXXVIII, 1897

Above: Puddling furnace used for the production of steel in the 19th century.
Unknown , "Household Encyclopaedia", (1906) .

"While unflinching in his professional life, privately he always retained his genial and loveable nature".

The Inst. C.E. Vol. CXXVIII, 1897

Southwark Street

The location of Kirkaldy's Testing Works on Southwark Street put it in the middle of an up-and-coming area in the mid-19th century. It was decided by the Metropolitan Board of Works in April 1856 to build a new street to stretch between the South Eastern Railway terminus at London Bridge Station and the West End. This made Southwark Street the first to be made by the Board, which was established in 1855. The street was completed in 1864, less than 2 years before David Kirkaldy opened his Works in The Grove just off Southwark Street.

Southwark Street swiftly became a centre of commerce, most probably due to it being adjoined to Borough High Street where Borough Market is located. In 1867, the Hop Exchange (designed by R.H. Moore) was opened on Southwark Street, which acted as a centre for hop trading for the brewing industry. This mercantile aspect to the street is probably what prompted Kirkaldy to move from his original location at The Grove to 99 Southwark Street.

Many of the buildings on Southwark Street follow the fashion of the time in terms of architecture. There is a significant trend of Italianate Romanesque and Gothic styles in the buildings, which is a theme that Kirkaldy's Works continues in its design.

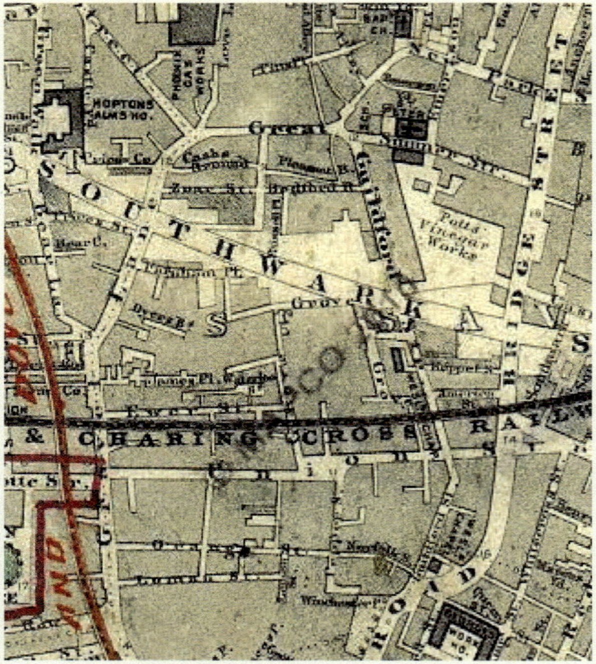

Above: Stanford's Library Map Of London And Its Suburbs 1864, enlarged to show The Grove off Southwark Street opened in 1865.

Below: engraving of Borough Market, c.1860.
Unknown, Illustrated London News, c.1860.

was signed with Messrs. Roberts on 8th April 1873 to conduct the building work, which was to be completed within seven months. Kirkaldy's testing machine was in place and fully functioning at 99 Southwark Street in 1874. The opening of 99 Southwark Street led to Kirkaldy being regarded as one of the foremost authorities on materials testing in the world. In 1885, he was elected as a member of the auspicious Institution of Civil Engineers, being proposed by J. W. Bazalgette and supported by a renowned group of engineers. He was also given his Honorary Freedom and Livery of the Worshipful Company of Turners on 6th December 1888, an honour by which he was said to be "deeply gratified".

Kirkaldy became renowned for his "sterling honesty" in his work to the point that he would stake everything he possessed as a guarantee of his responsibility for any work he conducted. As a result, he was present for all work conducted at his works. This passion for what he believed to be right, despite further confirming his reputation as a fearless and fierce individual, caused him great personal anxiety which dominated his life. His naturally sensitive disposition and obsessive tendencies were features throughout his life and only those closest to him knew how much this troubled him. David Kirkaldy's strongest point, that remains his legacy, is his desire that engineers should adhere to common standards. His vision, despite all opposition and hardship, is now fundamental in all engineering practices.

During the last two years of his life, his health deteriorated considerably, but Kirkaldy is reported to have been "cheerful and genial to that last" and still took great interest in knowing the details of experiments being conducted at his works. He died on Monday 25th January 1897 and is buried in Highgate Cemetery.

"A man of unique personality" who "saw the idea he once held almost alone become the commonplace of text-books".

'Engineering', 1897

"Imbued with a lofty sense of truth and honour, ardently striving to carry out his work impartially, he was almost recklessly regardless of consequences".

The Inst. C.E. Vol. CXXVIII, 1897

Above: David Kirkaldy
Unknown, The Engineer, 1897.

Right: the words David Kirkaldy had carved above the door of his testing works at 99 Southwark Street, which later became the family motto.

Above: William George Kirkaldy

Below: microscope awarded to William George

William George Kirkaldy 1862 – 1914

> *"In ordinary practise mechanical draughtsmen are no longer called upon to produce drawings ... such as characterised Mr. David Kirkaldy's superb sections of the SS Persia, exhibited in the Royal Academy over half a century ago, and now adorning the office of his famous son, Mr. W. G. Kirkaldy."*
>
> Machine Design, Construction and Drawing, Henry J. Spooner, C.E., Longmans Green, 1908,

David Kirkaldy's son, William George Kirkaldy, was born in Glasgow on 6th July 1862. Initially, he was privately educated until his attendance at University College School, London, which he left in 1880. In 1884, both son and father went into partnership, changing the firm title from David Kirkaldy to David Kirkaldy and Son. After this, all profits made by the firm were split one third to William George and two thirds to David. Within the Works, he was known to have the same zeal for research and experimentation as his father, engaging in a plethora of testing and experimental work until his death.

On his father's death in 1897, William George became the sole proprietor of the firm. Despite the majority of his work being associated with his father's testing Works, he was also a respected engineer and researcher in his own right. In 1899, in affiliation with Sir W. C. Roberts-Austen, a notable English metallurgist of the time, William George read a paper before The Institution of Civil Engineers (ICE) on "The Effects of Wear on Steel Rails". For this work, he was awarded a Telford premium, the highest prize awarded by the institute. Following this, he became closely involved in a number of committees and discussions within the institute, having already been made an Associate Member on the 6th December, 1898.

Having become an independent force in the world of materials research, William George became closely involved in the formation of National Physical Laboratory, opened in 1902 at Bushy House, Teddington. Originally intended as an extension of Kew Observatory, situated in Richmond, and under the instruction of the Royal Society, the laboratory has since become a hub for scientific investigation for all areas of practical Physics.

Left: Bushy House, Teddington, London.
Unknown, Popular Science Monthly, Volume 60, 1901/1902.

Below: Captain H. Riall Sankey, creator of the Sankey diagram.
Unknown, Institution of Mechanical Engineers, 1926.

In 1910, William George followed further in his industrious father's footsteps by developing a tensile impact machine. In collaboration with Captain H. Riall Sankey and Mr. Bertram Blount, he presented this research in a paper to the Institution of Civil Engineers on "Comparison of the Tensile, Impact-Tensile, and Repeated-Bending Methods of Testing Steel".

After the death of William George on 10th April 1914, his son, also David, was too young to run the works and so his widow Annie took over with a Dr. Gilbert H. Gulliver as manager, who remained at the works until his death in 1950. Despite the absence of a Kirkaldy at the helm, the Works continued to draw the clientele of some of the most exciting engineering projects of the time. For instance, the Works tested materials used in the construction of the original Empire stadium at Wembley, opened in 1923, proving the power wielded by the Kirkaldy brand.

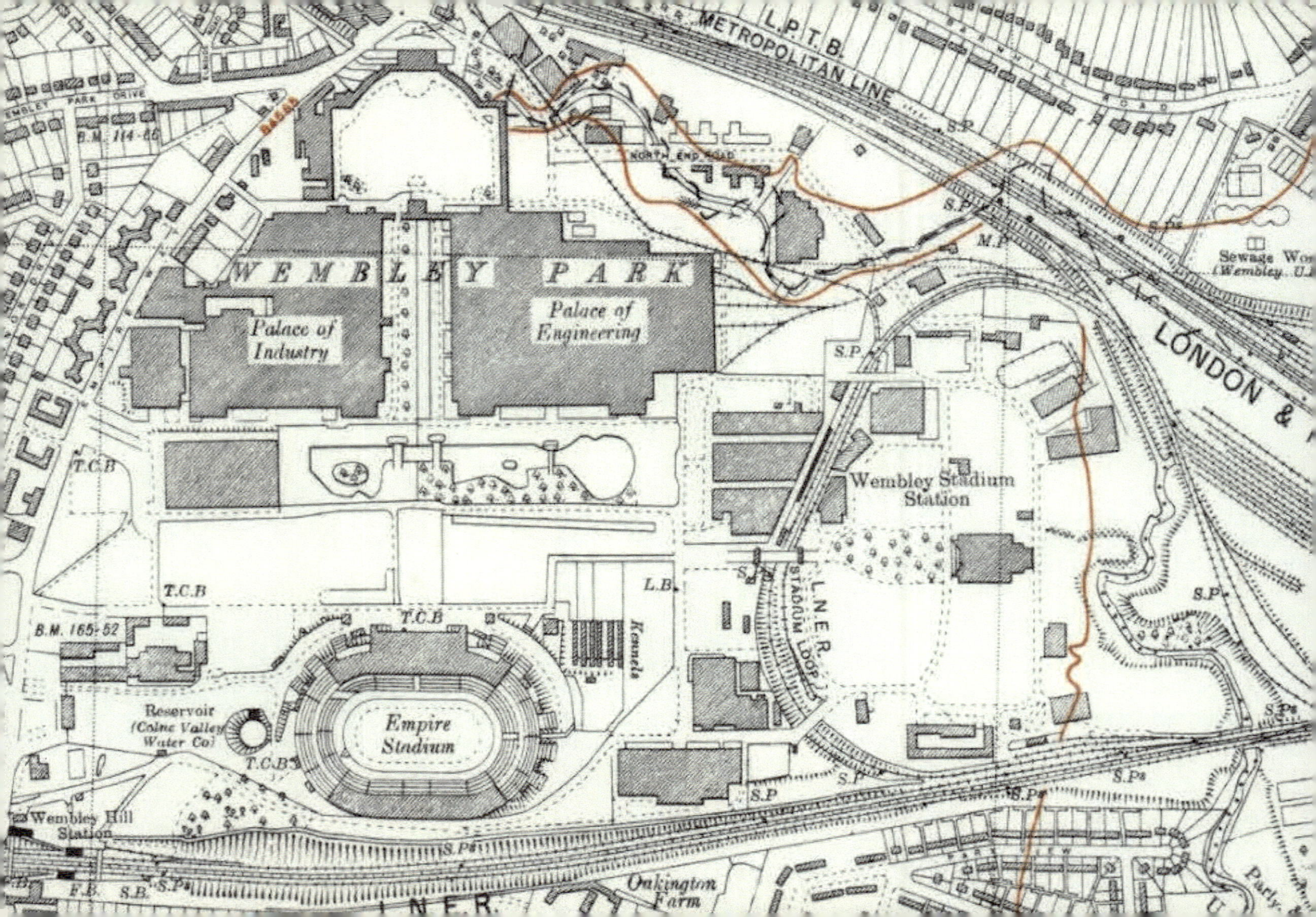

L.P.T.B.
METROPOLITAN LINE
WEMBLEY PARK DRIVE
B.M. 114·66
NORTH END ROAD
S.P.
M.P.
Sewage Wo
(Wembley U.
LONDON &
WEMBLEY PARK
Palace of Industry
Palace of Engineering
T.C.B
Wembley Stadium Station
L.B.
L.N.E.R.
STADIUM LOOP
Kennels
B.M. 165·52
Reservoir
(Colne Valley Water Co)
Empire Stadium
Wembley Hill Station
F.B.
S.B.
S.Ps
Oakington Farm
L.N.E.R.

David W. H. Kirkaldy 1910 – 1992

"But I am particularly fortunate for the chance meeting with David Kirkaldy (the founder's grandson) in Brighton and am most grateful to him not only for making the family papers available to me, but also for his friendship and encouragement in this project."

Dr. Denis Smith, Founder of the Kirkaldy Testing Museum, 'David Kirkaldy (1820 – 1897) and Engineering Materials Testing', 1980.

Above: David William Henry Kirkaldy.
Photograph courtesy of Chris Rule, GLIAS, date unknown.

David William Henry Kirkaldy was born in Hampstead in 1910, four years before his father's death in 1914. He was educated at Charterhouse before he went on to study at King's College Cambridge, where he took Mechanical Sciences. After gaining his Bachelor's degree, he began an apprenticeship at Vickers Ltd. River Don Works in Sheffield, which dealt with steel forging and casting. This gave him a firm footing for when he joined the family Works in September 1934, with his mother still acting as sole Proprietor. However, four years later, in September 1938, his mother, Annie Kirkaldy, died and David became Proprietor. Once more David Kirkaldy and Sons had a Kirkaldy at the reigns.

Left: Vickers Ltd. River Don Works, c. 1938.
Sheffield Libraries and Archives, www.picturesheffield.com, c. 1938.

Opposite: Ordnance Survey Map of Wembley Park area, London, 1938.
Image courtesy of Ordnance Survey, 1938.

David Kirkaldy continued to run the Works until his retirement to Brighton in April 1965, leaving no successor. He sold the works to Treharne and Davies, a firm of analysts and chemical consultants who already occupied the upper floors of 99 Southwark Street. However, despite the new occupancy, the testing works began to decline in the 1970s and eventually closed in 1974. So, there ended the 99 years of testing at 99 Southwark Street and 100 years of David Kirkaldy's Testing and Experimenting Works.

Below: an aerial view of the Works.

Rudyard Kipling's poem praising the efforts of the materials tester.

Hymn of Breaking Strain

Rudyard Kipling, 1935

The careful text-books measure
(Let all who build beware!)
The load, the shock, the pressure
Material can bear.
So, when the buckled girder
Lets down the grinding span,
The blame of loss, or murder,
Is laid upon the man.
Not of the Stuff - the Man!

But, in our daily dealing
With stone and steel, we find
The Gods have no such feeling
Of justice toward mankind.
To no set gauge they make us, -
For no laid course prepare -
And presently o'ertake us
With loads we cannot bear:
To merciless to bear.

The prudent text-books give it
In tables at the end -
The stress that shears a rivet
Or makes a tie-bar bend -
What traffic wrecks macadam -
What concrete should endure -
But we, poor Sons of Adam,
Have no such literature,
To warn us or make sure!

We hold all Earth to plunder -
All Time and Space as well -
Too wonder-stale to wonder
At each new miracle;
Till in the mid-illusion
Of Godhead 'neath our hand,
Falls multiple confusion
On all we did or planned -
The mighty works we planned.

We only of Creation
(Oh, luckier bridge and rail!)
Abide the twin-damnation -
To fail and know we fail.
Yet we - by which sole token
We know we once were Gods -
Take shame in being broken
However great the odds -
The Burden or the Odds.

Oh, veiled and secret Power
Whose paths we seek in vain,
Be with us in our hour
Of overthrow and pain;
That we - by which sure token
We know Thy ways are true -
In spite of being broken,
Because of being broken,
May rise and build anew.
Stand up and build anew!

Opposite; the ground floor of the works, 1885.

The Universal Testing Machine

Above: the Universal Testing Machine.

Design of the Testing Machine

> *"We have little doubt that civil, military, mechanical, and mining engineers, architects, ship and bridge builders, boiler makers, and others, will often find this testing apparatus advantageous in determining the best proportions of most structural details, as well as the materials, and variety of materials specially adapted to their different requirements. The machine may be employed for private as well as public purposes."*
>
> "The Engineer", 1865

The machine spans 47 feet 6 inches in length and 4 feet 3 ½ inches in width, the machine bed being comprised of four iron box-castings bolted together. At one of the far ends of the machine is mounted a hydraulic cylinder, which contains a ram of an effective area of 149 square inches and a stroke of about 3 feet. The machine operates in one direction, the ram within the cylinder operating a crosshead supported by two wheels, which moves as the ram slides under pressure, the cylinder remaining stationary. Four, square-threaded rods, each 5 ¾ inches in diameter and 36 feet long, are attached to the crosshead. These rods carry a carriage with a width of 11 feet, which can be moved along the rods by the means of two winding wheels attached to the carriage on either side. The mobility of the carriage allows the machine to accommodate specimens of different lengths. The carriage is also the means by which load is applied to a specimen. For instance, in a tension test, the test piece is gripped between two sets of wedge blocks; one set attached to the steelyard lever system and the other incorporated into the carriage.

> *"...it's not broken yet, and nothing leaves the machine till it's broken!"*
>
> *David Kirkaldy, quoted in The Engineer, 1897*

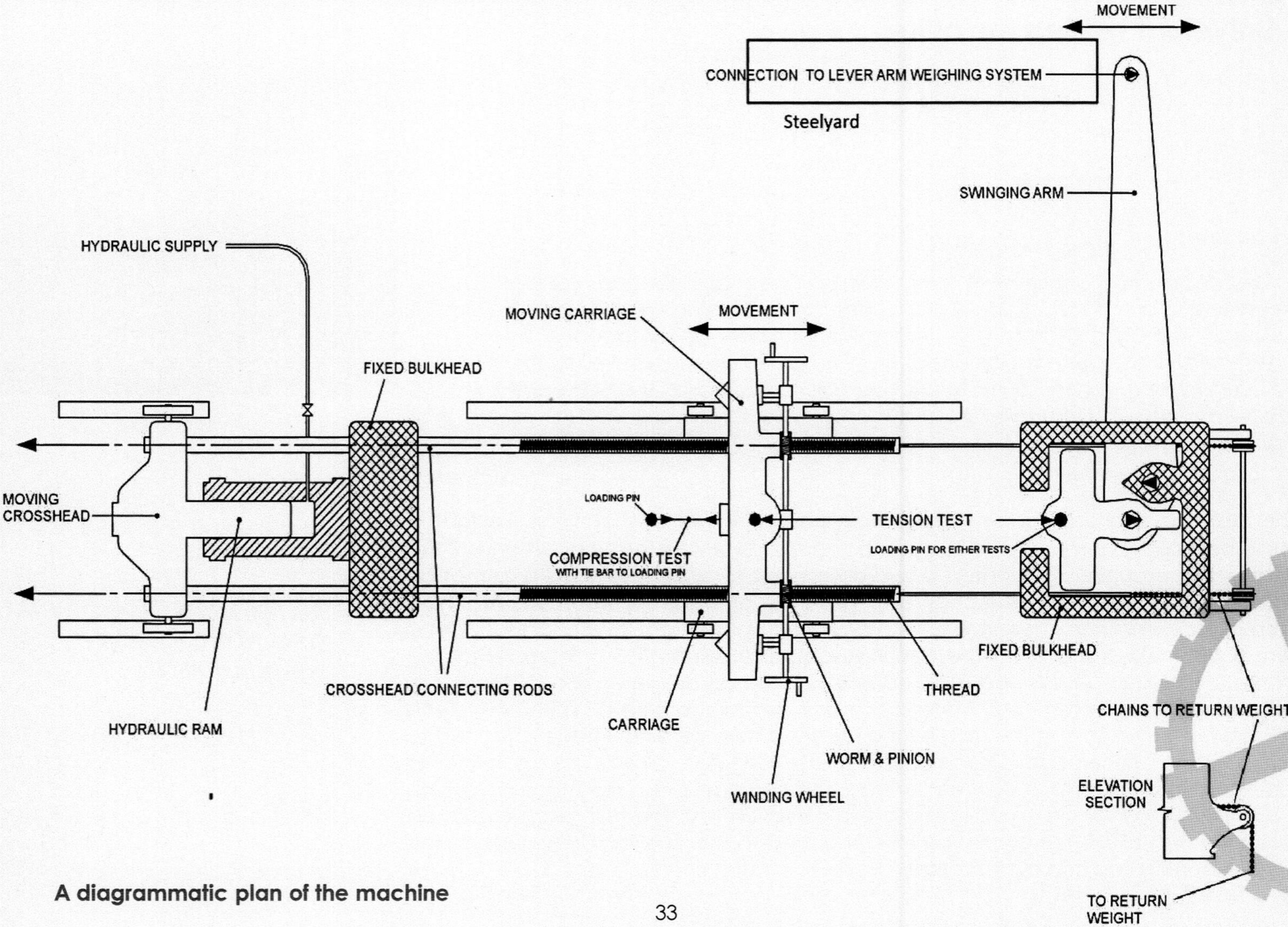

A diagrammatic plan of the machine

Schematically, the steelyard sits parallel to the machine, connected orthogonally to an extended lever. The horizontal lever transfers the load applied to the specimen to the steelyard through a leverage of 12:1. This means that the force communicated to the steelyard end of the lever is one twelfth of the actual specimen load. The large lever moves, in turn, a horizontal rod which, by use of a bell-crank, puts the counter balance in the steelyard in motion. The counter balance moves in a vertical plane, the lower balance being used for smaller loads and the higher for greater loads.

The scale graduations begin at 20 and are ¾ inches apart and represent actual leverage. In order to calculate the load applied to the specimen, it is necessary to multiply the steelyard reading and the total weight on the jockey. This jockey weight rides on the scale to allow fine adjustments to be made when at the steelyard and are added up to 1000 lbs.

The original David Kirkaldy said that his machine "was constructed for 1,000,000 lbs, but *that he* never had occasion to go above three hundred tons". By applying the full 1,000,000 lbs, the machine would be exerting a load of 446.4 tons on the specimen.

The high pressure water required to operate the machine was supplied by tapping into the London Hydraulic Power Company mains, which ran at 750 lbs/in² (approximately 52 bar) under London. The company was established in 1883 by an Act of Parliament, with the specific purpose of powering large workshop machinery. It was the successor of The Wharves and Warehouse Steam Power and Hydraulic Pressure Company, founded in 1871, and continued to operate until 1977. At its peak, the LHP Company covered much of London. The immense span of pipes were built over a period of half a century and eventually reached a length of 184 miles. This vast network meant that weekly in 1893, 6,500,000 gallons was pumped through the streets of London, reaching a peak of 32,000,000 gallons weekly in 1933. However, the supplied pressure would not have been sufficient to apply more than, approximately, 50 tons to a specimen. In order to reach the full capability of the machine, a hydraulic intensifier was added. For Kirkaldy's stated maximum load of 446.4 tons, a pressure of 6711 lbs/in² (approximately 463 bar) would have to a supplied, meaning a nine-fold intensification of pressure.

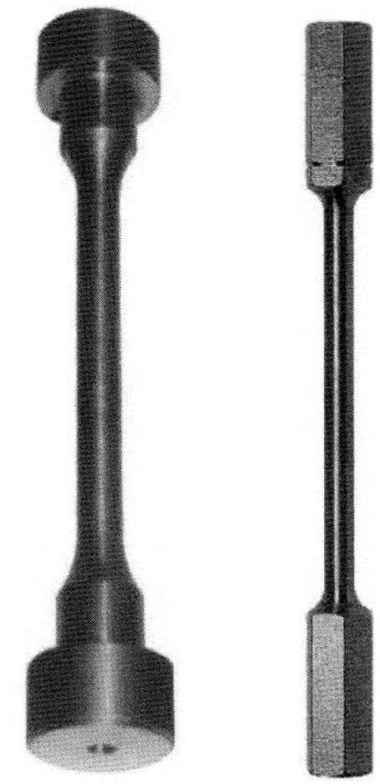

Above: steel tensile (left) and torsion (right) test specimens.
Photographs courtesy of Tecquipment Ltd.

Below: the premises of the LHP Company, Grosvenor Road, Pimlico.
Photograph courtesy of Charles Keeble.

PARK
CLERKENWELL
SHOREDITCH
SPITALFIELDS
BLOOMSBURY
WHITECHAPEL
CITY
SOHO
RIVER THAMES
HYDE PARK
SOUTHWARK
GREEN PARK
ST. JAMES'S PARK
PALACE GARDENS
BELGRAVIA
BERMONDSEY
NEWINGTON
WESTMINSTER
LAMBETH
PIMLICO

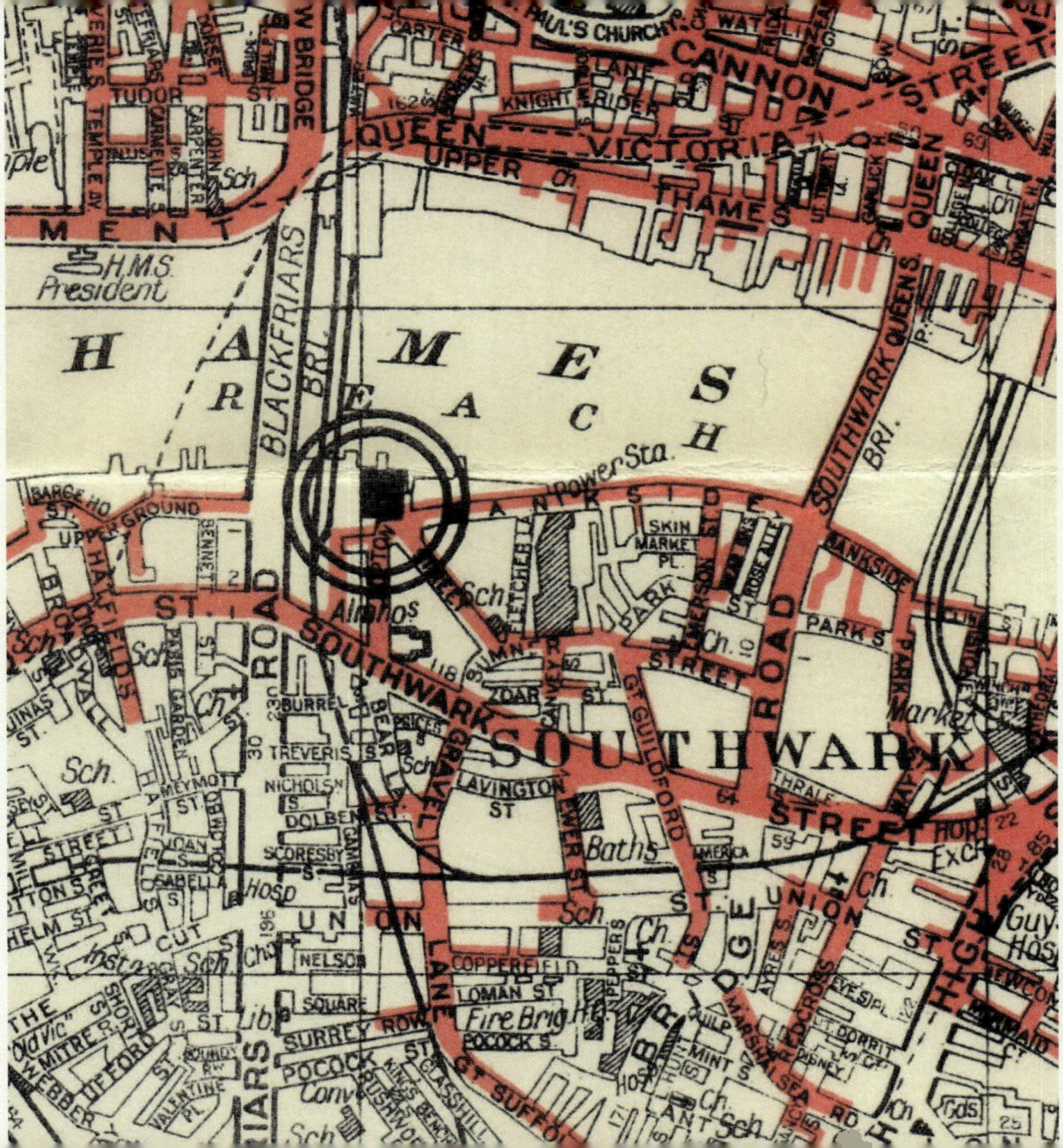

Enlarged map of the LHP Company's lines under running under Southwark.

Opposite: a map showing the placement of the LHP Company's lines under London.

Images courtesy of Subterranea Britannica.

The London Hydraulic Power Company

The LHP company mains were used to power a range of appliances throughout its operation. In addition to supplying large workshop machinery, it also was used to operate lifts, canes and theatre machinery. It famously provided power for the revolving stages at the London Palladium and the London Coliseum, the safety curtains at the Theatre Royal, the mechanism to lift the cinema organ at the Leicester Square Theatre and the Palm Court orchestra platform. Further to this, it was frequently used to power fire hydrants as well as providing a backup system for Tower Bridge.

The demand for a high pressure hydraulic supply began to decline in the early 20th century as electric power began to become more popular. As a result, the LHP Company began to install electric motors in their pumping stations in the place of the original steam engines.

The system was finally closed in June 1977, being made redundant in the wake of the firm establishment of electricity as a power source. Wapping Hydraulic Pumping Station (which was established in 1890) was the last of the five pumping station to close on 30th June 1977.

The pipe network was subsequently bought by Mercury Communications (a subsidiary of Cable & Wireless) which proceeded to use the pipes as telecommunications ducts.

This page: LHP mains in Piccadilly (c. 1930). The original iron pipes with oval flange spigot and faucet joints are being removed and replaced with new steel pipes using Victaulic joints.

Opposite page: engine room of the LHP Company's Grosvenor Road pumping station in Pimlico (c.1909). The triple expansion steam engines drive high pressure pumps.
Photographs courtesy of Subterranea Britannica.

Stress Analysis

Kirkaldy's Universal Testing machine is capable of four main classes of test on a specimen; tensile, compression, torsion and bending. Each of these tests occurs at a different point along the machine, with the addition or manipulation of various accessories. For example, tensile tests are conducted between the carriage and lever system whereas, compression tests occur between the carriage and cylinder bulkhead.

When analysing a specimen under tension (tensile test) or compression, two crucial calculations are those involving *stress* and *strain*. The stress upon a specimen is equivalent, dimensionally, to pressure (i.e. they are measured in the same units). It is, therefore, calculated as the force incident on the specimen per unit cross-sectional area. Units of measurement include: the Pascal (Pa), pounds per square inch (lbs/in^2) and bar (100,000 Pa).

$$Stress = \frac{Force}{Cross - Sectional\ Area} = \frac{F}{A}$$

The second vital quantity, strain, is a dimensionless quantity (i.e. it has no units), which is represented by the extended length per unit of the original length of the specimen.

$$Strain = \frac{Extension}{Original\ Length} = \frac{\Delta l}{l_o}$$

Once these quantities are calculated, the data can be presented in a stress – strain graph (see Figure 1). Representing the data in this manner allows key points in the extension of the specimen to be identified. In Figure 1, the straight part of the curve up to the "yield strength" describes elastic deformation of the specimen. This is

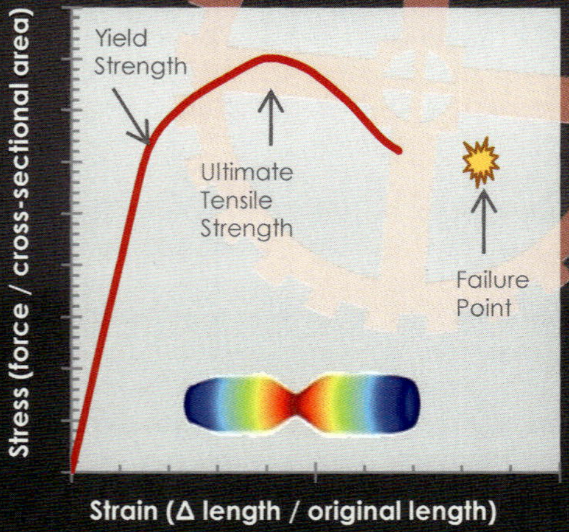

Figure 1: Stress – strain graph for a typical, metal specimen in tension. The vital stages in the test are labelled and a "necking" cylindrical specimen simulation is shown underneath (the colour in the image demonstrating stress concentration).

deformation which is reversible; once the load is removed, the specimen will return to its original shape or length. Under elastic deformation, the graph has a linear relationship, the stress and strain being proportionally related. This factor of proportionality is known as Young's Modulus or the Modulus of Elasticity:

$$Young's\ Modulus\ (Y) = \frac{Tensile\ Stress}{Strain} = \frac{F \cdot l_0}{A \cdot \Delta l}$$

This relationship leads to Hooke's Law, which sees the force exerted on a material as proportional to its extension, the constant of proportionality being formed of the Young's Modulus, original length and cross-sectional area:

$$Force = \left(\frac{Y \cdot A}{l_0}\right) \cdot \Delta l = k \cdot x$$

Any deformation past this point is *plastic deformation* and is permanent. Within the plastic part of the extension, the "ultimate tensile strength" represents the maximum stress that a material can withstand before failure. After this point, the material will fail under further strain to reach the "fracture point." Just before failure, it is common to see "necking" in a specimen in tension, appearing visibly as a narrowing near the centre where the stress is focused.

The analysis of a material undergoing compression is similar to that of one in tension, involving once more the creation of a stress – strain graph (see Figure 2). Again, a specimen will deform elastically under load prior to reaching its "yield strength". Once this point is past, the deformation is plastic and permanent. In parallel with the tensile strength, the specimen will also have an "ultimate compressive strength", representing the maximum compressive stress of material can take before fracture. Before failure, a specimen will undergo "barrelling" as its length is compressed, appearing towards as it bulges at its centre.

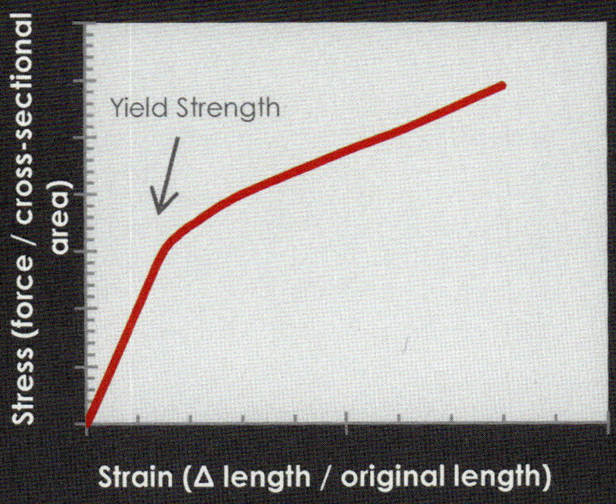

Figure 2: Stress – strain graph for a typical, metal specimen in a compression test. The yield strength of the material is labelled and a simulation of a "barrelling" cylindrical specimen is shown underneath (the colour indicates stress concentration).

An artwork by James Capper in tension at the Kirkaldy Testing Museum.
Photograph courtesy of Tommophoto.com, 2013

The Belgian Testing Machine

Far above and above: 't Arsenaal Mechelen

In 1877, the Belgian Government requested from Kirkaldy a replica of his, now famous, Universal Testing Machine. The order was placed once more with Greenwood and Batley:

> *"Kirkaldy's patent largest size Testing Machine, made on the same principle & from precisely the same models as the one used by Mr. Kirkaldy in his testing establishment at Southwark to test up to 500,000 kilos by drawing, crushing, bending, twisting, etc."*

The machine was installed in the Arsenal at Mechelen for the use of Le Ministre des Travaux Publics. The Belgian machine was delivered by the end of 1878. In charge of testing at Mechelen was M. Roussel, who personally visited the Works in Southwark frequently. In 1891, William George Kirkaldy visited the Belgian works noting:

> *"On reaching the Arsenal [I] had a very cordial welcome from Monsieur Roussel . . . Kirkaldy's Patent Testing Machine was found kept in the most perfect order, quite a treat to examine. Both M. Roussel and his assistant expressed themselves as highly pleased with the behaviour of the machine. They said it was as good as when new, the knife-edges being in splendid order. . . They are as proud of the machine I believe as we are."*

William George made further visits to Mechelen in the early 1900s, however, the testing machine has now been removed. Exactly when this was done or its whereabouts remain unknown.

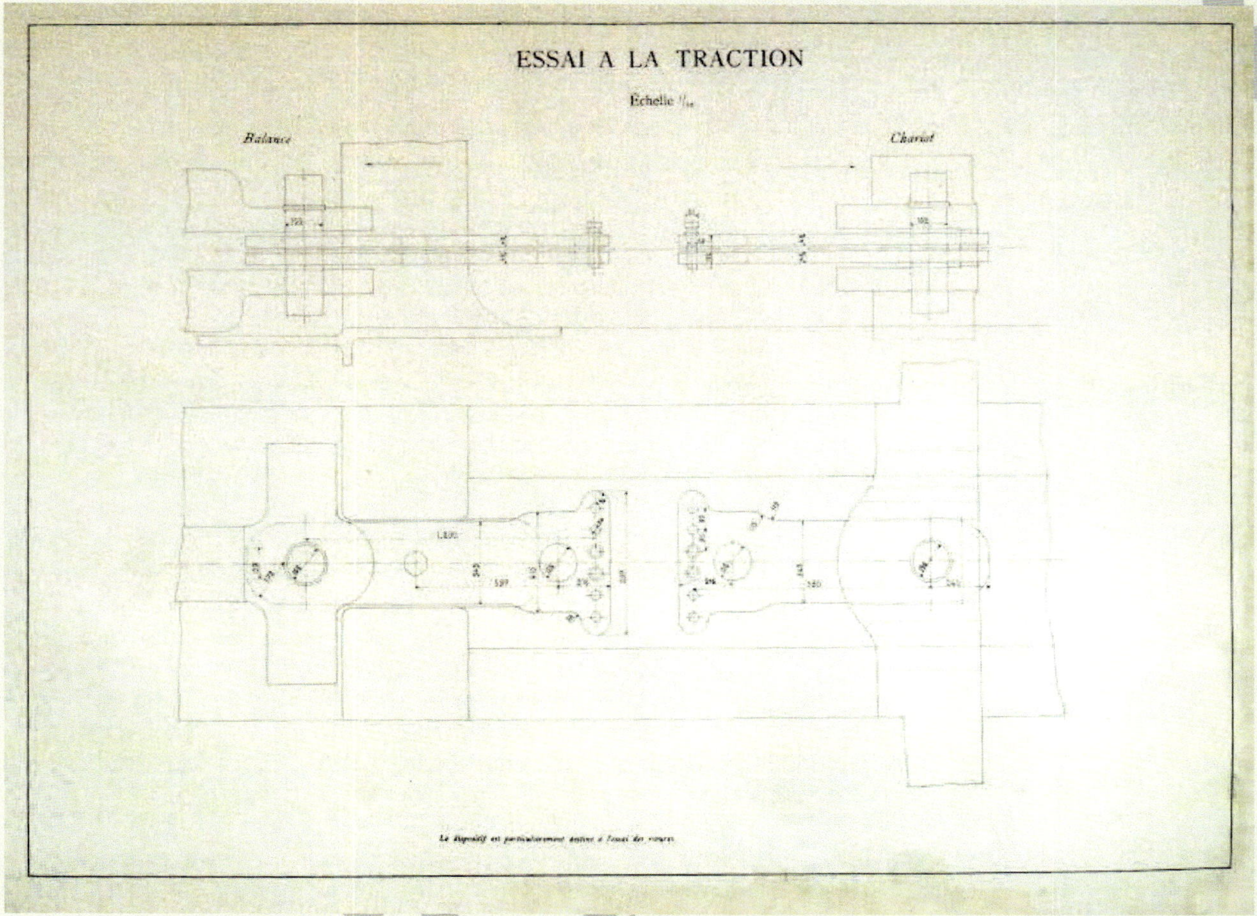

A diagrammatic plan of the Belgian machine

Above: Captain James Buchanan Eads, designer of the Eads Bridge.
Photograph by Mathew Brady, Library of Congress Prints and Photographs Division. Brady-Handy Photograph Collection, 1865-1880.

Below: sketch a of 19th century Mississippi steamboat used for the transportation of goods.
Unknown, Popular Science Monthly, Volume 12, 1876.

Eads Bridge, 1874

"The love of praise is, I believe, common to all men. And whether it be a frailty or a virtue, I plead no exception from its fascination. Yesterday, friends expressed to me their pleasure at the thought that my mind would be relieved after testing the bridge. But I felt no relief, because I felt no anxiety on the subject."
James B. Eads, 1874

During the early to mid-19th century, the transportation of goods across the Mississippi River was constricted to steamboat in the absence of a bridge capable of supporting vehicles. At the time, the design of the steamboat had many faults, which often led to their cargoes being sunk and lost. With the combination of sandbars, snags and exploding boilers, the loss of cargo was a common occurrence and retrieval from the bottom of the river proved difficult and costly. A bridge over the Mississippi at St. Louis was a long-standing idea, but its construction was delayed by the onset of the American Civil War (1861 – 1865), pushing the project well into the second part of the 19th century.

Following the conclusion the Civil War, the need for a bridge was becoming increasingly pertinent. The width of the river at St. Louis posed something of a barrier to goods transportation after the rise of the railways. In the absence of a bridge, goods had to be unloaded from the trains into ferries, which were unable to carry the freight in bulk. This added both time and expense to the transportation. As a result, shippers tended to re-route their cargo around St. Louis, in order to bypass the river crossing. Due to various political and technical issues, the project for a bridge was not formally accepted until 1867 when James Eads stepped into the fold.

In 1867, the plans for James Eads' radical design were accepted. His plans were particularly bold due to his choice of steel as his primary building material. The alloy was still relatively new, the creation of steel by the Bessemer process (a more modern form of steel) only having happened in 1855. This method, allowing large scale

Photomechanical print of the Eads Bridge under construction in the 1870s. At this stage, the ribs are completed and work on the roadways has begun
C.M. Woodward. St. Louis : G.I. Jones and Company, 1881, Plate XLV.

THE BRIDGE
AT ST. LOUIS.
Capt. James B. Eads, Chief Engineer
Col. Henry Flad, Asst. Chief Engineer

Opposite: chromolithograph published by Compton & Co. View of completed the Eads Bridge (St. Louis Bridge), with steamboats in Mississippi River, surrounded by eight views of stages of bridge construction. They are based on photographs taken in 1874 by R. Benecke, sections of pier and machinery and portrait of Capt. James B. Eads, chief engineer.

Image by St. Louis : Pub. by Compton & Co, 1870 – 1880.

Below: hand coloured lithograph of the Eads Bridge entitled " The great St. Louis Bridge - across the Mississippi River ".

New York : Published by Currier & Ives, 1874.

Publicity and Image

This image was published to celebrate the completion of the Eads Bridge. It had become an icon for St. Louis; a symbol of innovation and strength. Many strange and wonderful tests were carried out on Ead's symbolic creation in order to exemplify this iconic strength. It is reported on 14th June 1874 that John Robinson led a "test elephant" across the bridge as proof of its sturdiness. There was great excitement from the onlookers in the crowd who watched this elephant, seconded from a travelling circus, making its steady way towards the Eads Bridge. It was a commonly held belief of the time that animals, such as elephants, had such acute instincts and heightened sense of danger that they would not cross a structure that was unsafe. Thankfully, the elephant crossed without any apparent qualms, proving its robustness to the crowd. Further to this, two weeks later, a convoy of 14 steam locomotives were driven back and forth across the length of the bridge. Once again, the Eads Bridge passed muster, demonstrating its soundness of design.

On the bridge's official opening day on 4th July 1874, there was a parade which extended 15 miles through St. Louis.

Above: photograph of the Eads Bridge entitled "The Magnificent Eads Bridge, St. Louis, Mo." *Robert N. Dennis collection of stereoscopic views, 1899.*

Below: drawing of the Eads Bridge, 1875. *Artwork by Camille N. Dry, Pictorial St. Louis, the great metropolis of the Mississippi valley; a topographical survey drawn in perspective A.D. 1875.*

production of steel at low cost, was to completely replace wrought iron in the following years. The Eads Bridge, therefore, became the first major structure to have true steel as its primary building material and, consequently, was to start a revolution in building materials. The actual design of the bridge was also fairly daring, not using the popular construction features of the time. Instead, Eads' design appeared to echo the methods exercised by the ancient Romans; more specifically, the Roman arch. Each of these arches was comprised of four ribs, arranged in parallel pairs, supporting a double deck roadway. Connecting these arches and securing them to the riverbed are four stone towers; two on each shore and two in the middle of the river. The unconventional nature of Eads' design, and the fact that he had not designed a bridge before, led to many declaring him mad and urging him to seek more experienced help.

It has been recorded that several samples to be used in the steel superstructure of the bridge were sent to Kirkaldy's Works for testing. The transaction is recorded by John A. Kouwenhoven in his notes concerning the construction of the bridge:

> *"(Eads) had furnished thirty-two samples of (chrome steel) to David Kirkaldy, the pioneer of steel testing in Great Britain, in 1868…He provided Kirkaldy with more samples to test on April 7, 1869…"*

The construction of the bridge was plagued with problems ranging from workers suffering from the "bends" during the sinking of the caissons to spring floods to a tornado which struck in 1871, crippling the superstructure of the eastern abutment. These problems led to multiple injuries and fatalities amongst the workforce.

On completion, the bridge contained 4,780,000 pounds of steel and 6,313,550 pounds of wrought iron. It was comprised of three arches of steel, each spanning more than 500 feet in length. The upper deck was for the use of carriages and horsemen, whereas the lower deck was to support a railroad. Eads had "overdesigned" his bridge in order to exaggerate its apparent strength to the public. To further confirm its strength, in 1874 he conducted a highly publicised test of its strength by running 14 large locomotives out onto the bridge. To this day the bridge can carry 5,000 pounds per linear foot. Two days after this test, on 4th July 1874, the bridge was formally opened as the largest and first steel arch bridge in history.

The Tay Bridge Disaster, 1879
By Colin Jenkins

"Oh! ill-fated Bridge of the Silv'ry Tay"
The Tay Bridge Disaster, William McGonagall, 1880

Above: engraving of Sir Thomas Bouch, designer of the Tay Bridge.
Unknown, Illustrated London News, 1877.

The River Tay estuary at Dundee is around 2 miles wide, so in the 19th century it was a major barrier separating Fife and the growing city of Dundee. The North British Railway Company proposed to build a bridge over the estuary and obtained a parliamentary act for its construction. The first contract was let in 1871 and construction took place through the 1870s. The bridge was opened on 31st May 1878 and carried Queen Victoria on her way to Balmoral.

The designer of the Bridge was Sir Thomas Bouch (1822-1880) and the structure he designed had to be wide enough between spans and high enough not to impede river navigation. It was built as a series of 85 spans mounted mainly on cast iron columns with the central section from piers 28 to 41 raised to a greater height than the rest as can be seen in the picture. In these central high girders trains travelled between the girders rather than above them.

The night of Sunday 28th December 1879 was a particularly stormy one when the Edinburgh to Dundee train came up to the southern signal box at 7:13pm to obtain the token for the bridge section. It then proceeded across but the signalman reported the tail lamps disappeared and never appeared again.

Below: photograph of the Tay Bridge.
Photograph by Valentine's of Dundee, 1878/1879.

Illustration of the Tay Bridge disaster, printed in the Illustrated London News, 1880.
W J F, Illustrated London News, 1880.

Illustration of the Tay Bridge disaster, printed in The Graphic, 1880.
The Graphic: An Illustrated Weekly Newspaper, 1880.

The train had been blown against the girders and the whole central structure with the train had fallen 90 feet into the river. There were no survivors. From the inquiry evidence, recovery boats only reached the site several hours later. Over the following days divers were sent down to examine the wreckage and retrieve parts of the bridge, including the locomotive which was later rebuilt and returned to service.

A Board of Trade inquiry into the disaster was opened on Saturday 3rd January 1880 under 3 commissioners. It sat until May and examined around 120 witnesses. David Kirkaldy was never called as a witness but his name appears 157 times in the evidence and reports, according to a handwritten sheet inserted into David Kirkaldy's copy held by the Museum. As the leading expert on the strength and failure of materials at that time, he was asked to undertake tests on the components recovered from the riverbed. He tested the strength of the bolts which fixed the cast iron columns to the pier bases and the lateral tie bars of the columns. These were tested to destruction but during the tests the extension under load was recorded. The force used starting at 10,000 lbs. then increasing by increments of 2,000 lbs. These connecting components were clearly one focus of the enquiry.

Below: photograph of the salvage of parts of the Tay Bridge.
Board of Trade, National Library of Scotland, 1879.

Below left: photograph of the collapsed Tay Bridge.
Photograph by Valentines' of Dundee, date unknown.

The other was the strength of the cast iron columns themselves which were cast at a foundry setup specifically for the bridge construction. A number of witnesses had attested to poor quality casting leading to uneven thicknesses and voids which were filled with a putty-like filler. For the tests undertaken by Kirkaldy, pieces were cut at random from the fallen columns, these being 1 inch broad by 2 inches deep and had a 20 inch bearing. They were subject to tensile tests where a mean strength of 9.14 tons per square inch is quoted. In crushing tests on the other hand, the mean was 35 tons per square inch showing the great difference between cast iron in tension and compression

The outcome of the inquiry was that the fall of the bridge was likely to be due to failures in the cross bracing but exacerbated by design deficiencies, poor workmanship in the casting of the columns and insufficient maintenance. However one inspector produced a separate report which agreed with the technical conclusions but laying the blame for the failure on Sir Thomas Bouch, something which the other two inspectors were reluctant to do.

The test results were not presented to the inquiry by David Kirkaldy, instead these were interpreted by others, in particular the civil engineer Mr Henry Law who undertook the technical analysis relating to the structure. In David's copy of the report are his marginal comments, often in red ink. Where Mr Law states that he would convert the bolt strength results to a stress using the thread area which has a smaller cross section, hence implying the strength of the material is greater Kirkaldy has written " Only a knave or fool would do so". Slightly later the printed reports has "The way in which Mr Kirkaldy reduces his experiment is clearly erroneous and "I fancy he has fallen into some error". Beside such comments David has scrawled "A lie". It is clear that someone of his professional integrity would have been offended by seeing comments like that in print.

In conclusion, Sir Thomas Bouch died in the same year but one of the commissioners Henry Barlow went on to design the mighty Forth Bridge. This was built in steel but is engineering on a different scale to the lightweight Tay Bridge structure and clearly a response to its weaknesses.

Far above: the steam engine salvaged from the river Tay.
Board of Trade, 1879 – 1880. Creative Commons Attribution 4.0 .

Above: salvage operations in progress in the Firth of Tay.
Board of Trade, National Library of Scotland, 1879.

1414 FALLEN GIRDERS BEACHED, TAY BRIDGE VALENTINES SERIES

This page and opposite: photographs of the Tay Bridge after the disaster.
Photographs by Valentine's of Dundee, date unknown

Valentine's of Dundee

Valentine's of Dundee, headed by James Valentine, was a Scottish photography firm, which achieved international fame for producing picture postcards. Much of the company's fame came about after the Tay Bridge disaster, when they were commissioned by the Court of Inquiry to photograph the remains of the collapsed bridge. In total, they took over 50 high quality images, which were used in the court as an aid to witnesses when giving their testimonies.

These photographs were sold as postcards across the world, contributing to Valentine's of Dundee's subsequent global status.

Right: photograph of James Valentine, c.1850.
Photograph by Thomas Rodger, c.1950.

William McGonagall's poem about the collapse of the Tay Bridge. It was famously known to be of such low quality as to be comical

The Tay Bridge Disaster

William McGonagall, 1880

Beautiful Railway Bridge of the Silv'ry Tay!
Alas! I am very sorry to say
That ninety lives have been taken away
On the last Sabbath day of 1879,
Which will be remember'd for a very long time.

'Twas about seven o'clock at night,
And the wind it blew with all its might,
And the rain came pouring down,
And the dark clouds seem'd to frown,
And the Demon of the air seem'd to say-
"I'll blow down the Bridge of Tay."

When the train left Edinburgh
The passengers' hearts were light and felt no sorrow,
But Boreas blew a terrific gale,
Which made their hearts for to quail,
And many of the passengers with fear did say-
"I hope God will send us safe across the Bridge of Tay."

But when the train came near to Wormit Bay,
Boreas he did loud and angry bray,
And shook the central girders of the Bridge of Tay
On the last Sabbath day of 1879,
Which will be remember'd for a very long time.

So the train sped on with all its might,
And Bonnie Dundee soon hove in sight,
And the passengers' hearts felt light,
Thinking they would enjoy themselves on the New Year,
With their friends at home they lov'd most dear,
And wish them all a happy New Year.

So the train mov'd slowly along the Bridge of Tay,
Until it was about midway,
Then the central girders with a crash gave way,
And down went the train and passengers into the Tay!
The Storm Fiend did loudly bray,
Because ninety lives had been taken away,
On the last Sabbath day of 1879,
Which will be remember'd for a very long time.

As soon as the catastrophe came to be known
The alarm from mouth to mouth was blown,
And the cry rang out all o'er the town,
Good Heavens! the Tay Bridge is blown down,
And a passenger train from Edinburgh,
Which fill'd all the peoples hearts with sorrow,
And made them for to turn pale,
Because none of the passengers were sav'd to tell the tale
How the disaster happen'd on the last Sabbath day of 1879,
Which will be remember'd for a very long time.

It must have been an awful sight,
To witness in the dusky moonlight,
While the Storm Fiend did laugh, and angry did bray,
Along the Railway Bridge of the Silv'ry Tay,
Oh! ill-fated Bridge of the Silv'ry Tay,
I must now conclude my lay
By telling the world fearlessly without the least dismay,
That your central girders would not have given way,
At least many sensible men do say,
Had they been supported on each side with buttresses,
At least many sensible men confesses,
For the stronger we our houses do build,
The less chance we have of being killed.

Illustration of the Tay Bridge disaster, printed in The Graphic, 1880.
The Graphic: An Illustrated Weekly Newspaper, 1880.

Hammersmith Bridge, 1887, and Albert Bridge, 1873

By Hugh MacGillivray

> *"A suspension bridge, with carriage road, spans the river at this point, and was for many years a favourite and cheap grand stand on the University Boat-race day .... It is now (1885) being rebuilt."*
>
> Dickens's Dictionary of the Thames (1885)

Hammersmith Bridge is an elegant suspension bridge over the River Thames designed by the famous Victorian engineer Sir Joseph Bazalgette. It replaced an earlier suspension bridge of 1827 by William Tierney Clarke but rests on the original pie foundations. The bridge was built by Dixon, Appleby & Thorne and was opened by the Prince of Wales on 11 June 1887. It is 700 feet (210 m) long and 43 feet (13 m) wide and cost £82,117 to build. The design allows a clear span of 400ft (122m) over the river, and a height above the water of 12ft (4m).

Above: photograph of Sir Joseph Bazalgette. *Photograph by Lock & Whitfield, National Portrait Gallery, London, 1877.*

Below: engraving of the first Hammersmith Bridge, 1827. Project Gutenberg, etext 12595, 1827.

Bazalgette's design uses flat steel suspension links rather than cables. The Institution of Civil Engineers has kindly supplied a copy of a page from The Engineer of 22nd April 1887, showing part of Bazalgette's drawing of these links, which are still in use today The flat chain links were manufactured by John Brown & Co of Sheffield. Each was forged by rolling from an ingot of mild steel, that is steel with a low carbon content, into a bloom of the specified shape. No re-heating or hammering was used.

The following quotation is from Hammersmith Bridge by Charles Hailstone, published by the Barnes & Mortlake and Fulham & Hammersmith Historical Societies in 1987.

> *"Before acceptance [the links were] subjected to 30 tons per square inch by Messrs. Kirkaldy of Southwark."*

There were a total of about 2200 links, each weighing about 650lb/300kg.

> *"The suspension chains on the bridge consist of four sets, two to each side. Each set consists alternately of eight links of 9in by 1 1/8in, and nine links of 9in by 1in, with an average length in excess of 16 ft. At the time they were the largest ever manufactured. Each set of links was bored at the same temperature, for the pinholes, so that the 6in diameter steel pins prepared by the Thames Iron Works could pass through them without being driven. The securing nuts are 9in across."*

We have always known that David Kirkaldy was involved in testing parts of Bazalgette's bridge, but a recent close examination of some 19th century photographs has revealed much more information. These four photographs are views of the Testing Works and have been on show in the Museum at 99 Southwark St since the 1980's. They were removed from their frame and scanned at high resolution for use in our fund-raising appeal brochure. This has allowed a close inspection of the detail in the pictures for the first time!

The ground floor picture (page 31) shows Kirkaldy's 440ton testing machine polished or oiled to look its best for the photographer. It is set-up for tensile testing but without the wedge-grips boxes we normally use now. The four short links used with the wedge-boxes are lying on the floor in two pairs. The ram is fully entered into the cylinder – something we cannot do now. The Greenwood & Batley nameplate can just be read as "Kirkaldy's Patent Testing Mac[hine]." Four long flat links are stacked on the floor, which are possibly from Hammersmith Bridge. This would date the photographs to around 1886-1887. Just below the machine crosshead is the broken end of a similar link.

The first floor workshop shows an incredible level of detail, and would repay scrutiny

2112—Hammersmith Bridge, London, England.

Above: photograph of the current Hammersmith Bridge, 1896.
Keystone View Company, Library of Congress Prints and Photographs Division Washington, 1896.

Below: photograph of Hammersmith Bridge, 2008.
Photograph by Flavio Ferrari, 2008.

Above: a repaired chain link from the IRA bomb attack in 1939.

Below: Hammersmith Bridge chain links.

by an expert in the development of machine tools (page 89). Test samples are being manufactured from components sent in by Kirkaldy's clients. Centre leaning against a bench are 2 broken ends from flat links which, again, could be from Hammersmith Bridge. Behind them to the right is a row of what look like 6-inch pins from the bridge.

The second floor store-room has been photographed from the east end, looking west towards the entrance and stairs (page 12). Front and centre are two long flat chain-links, also possibly from Hammersmith Bridge; they are both broken centrally by straight fractures with only limited ductile necking visible. Behind them and to their left are broken parts of similar links, two showing complex fractures in the pin-hole itself. At the very back against the right-hand windows are many parts of broken links.

The link dimensions of 9in. x 1.125in tested at 30 tons/in sq would indicate a load of 303.75 tons, so clearly Kirkaldy was using the upper measuring scale on his machine regularly at this period. Such a load would be at or beyond the yield stress for mild steel, so this was a true proof test which would have identified any inferior material or significant defects in the steel. The sound of a link failing suddenly at 300 tons would have been heard all along Southwark Street!

The chain links survived an IRA bomb attack in 1939 which attempted to destroy the bridge. Two suspension rods were broken and one link from the upper chain on the upstream side was "thrown from its set into the roadway. On the lower chain, on which the suitcase bomb had lain, one link was bent round into a half circle towards the roadway, leaving the eight other links in the set buckled." The broken chain was repaired by R H Green & Silley Weir Ltd using a massive brace which can still be seen.

The bridge girders, decking and suspension rods have been replaced several times, but the original chain links are still in place. Following damage by heavy vehicles, the weight limit has been progressively reduced from to 15T to 12T and now 7.5T, although one 15T bus is allowed to cross the bridge at a time.

Following detailed reading by HM in 2016 of W G Kirkaldy's 1891 book Strength and Properties of Materials with a description of the System of Testing, the following nformation can be added:

On page 66, Experiments upon Steel Bridge-Links Reports N, O, and P, WGK describes esting of links for both the Albert and Hammersmith Bridges across the Thames. The original Albert Bridge design by R M Ordish had been authorised in 1864 and used a combination of parabolic cables and inclined stays. It was not completed until 1873 and from its opening was subject to severe vibration and perceived to be structurally weak.

Table N exhaustively details the results from 5 links tested in 1888 for Edward Bazalgette for the strengthening works. The links were described as having rolled edges, and were manufactured John Brown & Co, Atlas Works, Sheffield. Three of the links were 6.20 x 1.20in, and two 6.20 x 1.0in in section, with initial lengths between pin holes of 263 to 275in [about 22ft]. All five links were loaded to failure at between 185 and 248tons, with maximum stress between 30.5 and 33.0 tons/in2, and extensions of 13 to 15%.

An examination of the bridge on 13th March 2016 shows that the main catenary chain is made of links each approximately 20ft long. Their shape is similar to the diagram in Table N. All the links are 6.20in wide and they are connected in sets alternately 5 x 1.0in thick then 4 x 1.2in thick. These appear to be the original links corresponding to the design tested by Kirkaldy – no marking could be seen on the main links, but one of the drop links to the bridge deck was stamped BEST or WEST. Note that these link dimensions are almost identical to the extension links which form part of David Kirkaldy's 1865 440ton capacity testing machine!

W G Kirkaldy's book gives more detailed information on the testing programme performed on the links for Hammersmith Bridge. WGK states that the makers, John Brown & Co, "paid great attention to the uniformity and regularity in manufacture of the steel, with very marked success.......once the process was in thorough working order." Sadly he does not give details of the steel making procedure. One preliminary link was tested at full size i.e. 9.0in x 1.0in and this broke at one end, through the curved part of the neck. The failure load is not stated, but was probably between 275-300tons.

Above top: an etching of Chelsea Embankment and the Albert Bridge under construction, 1873. *Unknown, Illustrated London News, 1873.*

Above: photograph of Albert Bridge. *Photograph by Chris Whippet, Geograph project, Creative Commons Attribution-ShareAlike 2.0 license, 2009.*

Above: an etching of the Albert Bridge, 1873.
Illustrated London News, 1873.

Below: photograph of the Hammersmith Bridge.
Photograph by Dr. Neil Clifton, Geograph project, Creative Commons Attribution-Share Alike 2.0 Generic, license, 1982.

Table O exhaustively details the results from a batch of 40 trial links tested during the summer and autumn of 1886. The parallel section of each link was reduced by planning on one face only, from 1.0in to 0.7/0.8in thick. This was presumably to ensure that fracture occurred in the parallel section, not at the ends. The average yield stress was 17.1tons/in2, and the ultimate stress 30.0tons/in2. However, most of the broken surfaces showed significant areas [50 – 80%] of granularity, what is now called cleavage fracture, with at least four links showing 100% cleavage and one of these failing at a distinct flaw.

Table P details tests on a second batch of 40 trial links made during the same period. In this case the width of the parallel section was reduced by shaping presumably symmetrically though this is not stated, from 9in to 0.6/0.8in. The average yield stress was 17.3tons/in2, and the ultimate stress 29.8tons/in2. Of this batch, only seven links showed more than 50% granularity, but again four showed 100% cleavage; there were no failures caused by flaws. Either this was a better batch of steel, or reducing the thickness of the links and hence one component of triaxial stress, had discouraged cleavage fracture.

WGK states that the procedure for testing the production links involved selecting two from each batch of 50. It is not stated, but these were probably tested as received and not machined down. If both samples were unsatisfactory the whole batch was rejected. If only one was below standard, a third link was tested, and if this was satisfactory, the batch was accepted. This procedure apparently worked well, but WGK does not say if any batches were actually rejected. For a total of 2200 production links, this gives 44 batches at 2 links per batch, so at least 88 links were broken. Adding the trials, this implies a grand total of at least 169 link tests to failure using Kirkaldy's machine.

"...may the bridge tend to communicate prosperity and wealth from one end of the island to the other..."

Duke of Sussex, opening ceremony of the first Hammersmith Bridge, 1825.

Sydney Harbour Bridge, 1932

> *"The Bridge has dwarfed the city and humbled the North Shore. It sets the scale for a new era and a modern Babylon. Old Sydney is now but a memory"*
>
> Lionel Lindsay, Australian artist.

The idea of bridge adjoining the Northern and Southern harbours in Sydney had been a long standing concept prior to the construction of Sydney Harbour Bridge in 1932. It was first proposed by the Government architect Francis Greenway in 1815, who believed such a bridge would:

> *"...give an idea of strength and magnificence that would reflect credit and glory on the colony and the Mother Country."*
>
> Letter to *The Australian*, 1825

Following Greenway's suggestion were a multitude of proposals throughout the nineteenth century. These proposals were diverse, including a floating bridge in 1840, a truss bridge in 1879 and a high-level bridge in 1880. The turn of the century signalled the Australian Government's formal invitation of design submissions for the bridge. However, none of these designs were seen as suitable and the project was abandoned.

The project was not to be revived until after the commencement of the First World War by a senior engineer for the Department of Public Works J.J.C. Bradfield. Bradfield advocated the building of a cantilever overpass, and in 1916 the Legislative Assembly of Australia passed the bill for the construction of a cantilever bridge between Dawes Point and McMahons Point. The construction, however was once again put aside following the Legislative Council's rejection of the proposal on favour of the funding the war efforts.

Above: John Jacob "Job" Crew Bradfield, designer of Sydney Harbour Bridge.
Unknown, National Library of Australia, date unknown.
© State of New South Wales (Service NSW) 2016

Below: stamp celebrating the opening of the Sydney Harbour Bridge, 1932.
Photograph by Steve Strummer, 2010.

Above: from left to right: Chancellor Sir. William Cullen, his wife and Dr. J.J.C. Bradfield, the chief engineer and designer of the Sydney Harbour Bridge. Taken during construction. *Dr. J.J.C. Bradfield and W.H. Lush, University of Sydney Library's exhibition on the construction of the Sydney Harbour Bridge, date unknown.*

It was not until 1922, when Bradfield managed to gain overseas investment anc advice, that the Act to build Sydney Harbour Bridge was passed as a high-leve cantilever bridge (arch bridge) between Dawes Point and Milson's Point. Actua construction commenced on 28th July 1923 and the contract was later given to the English firm Dorman Long & Co of Middlesbrough on 24th March 1924. The constructior was undertaken by 1,400 workers and took eight year to complete at a total cost o $4.2 million.

The Sydney Harbour Bridge is composed of two major features; the steel arch anc the concrete pylons. The arch comprises two 28-panel arch trusses, their height tapering towards the centre (from 187ft at the ends to 59ft at the centre). It has a tota span of 1654ft and a maximum height of 440ft, allowing the bridge to hold its place a the sixth longest spanning-arch bridge and tallest steel arch bridge in the world. As c single piece, the arch weighs approximately 38,400 tons, 79% of which was importec from England. This particular connection is perhaps why Kirkaldy's Testing Works wa sourced to conduct strength tests on the steel used in the framework. Despite c

Far left: erection of hanger with special cradle. *Dr. J.J.C. Bradfield and W.H. Lush, University of Sydney Library's exhibition on the construction of the Sydney Harbour Bridge, date unknown.*

Left: southern anchorage system. *Dr. J.J.C. Bradfield and W.H. Lush, University of Sydney Library's exhibition on the construction of the Sydney Harbour Bridge, date unknown.*

distinct lack of records of the transaction, it is likely given this information that the steel was passed to Kirkaldy's Works before being sent on to the site of construction in Australia. Once this steel was formed into girders at one of two on-site workshops at Milson's Point, it was held together in the framework by six million Australian made driven rivets (the process of welding having not been sufficiently developed at this time). At each end of this vast network of steel girder and rivets are two 292ft high concrete pylons, each faced with granite. These pylons, comprising over 3 million cubic feet of concrete and over 600,000 cubic feet of locally sourced granite, serve no structural purpose with the exception of the abutments at the base which support the load of bridge. They were not originally a part if Bradfield's designs but were added later for mainly aesthetic reasons.

Above: workers on the Sydney Harbour Bridge.
Dr. J.J.C. Bradfield and W.H. Lush, University of Sydney Library's exhibition on the construction of the Sydney Harbour Bridge, date unknown.

In 20th August 1930, the two sides of the arch finally met in the centre meaning marking the bridge becoming self-supporting. The event was marked by the flying of the Australian and British flags from the jibs of the creeper cranes. The pylons were later built on top of the abutment tower and the bridge was load tested, the four railway tracks having been completed, with 96 steam locomotives in February of 1932. Today the bridge has only two rail tracks, the remaining eight being for traffic.

Below: poster for the opening of the Sydney Harbour Bridge, 1932.

Premier Jack Lang opened Sydney Harbour Bridge on 19th March 1932, witnessed by between 300,000 – 1,000,000 onlookers. Despite Australia being deep within the depths of the Great Depression, the celebrations were extensive, including decorated floats, marching, a gun salute and passenger ships running under the bridge alongside a Venetian Carnival. For this reason, and the fact that is kept so many workers employed, the bridge was regards as a symbol of Australia's resilience against the Great Depression, becoming known as "the Iron Lung". Today, the bridge still stands as an icon of Australia and its triumph through hardship.

"The Coathanger"
nickname of the bridge given by the Sydney locals

3 8 28
MANCHESTER

Collection of photos taken by Dr. Bradfield and his colleagues during the construction of the Sydney Harbour Bridge.

This page – Top left: creeper crane prior to dismantling.
Bottom left: southern abutment tower.
Above: bridge roadway.
Opposite – Top left: making connection of centre hanger with lower chord.
Bottom left: lower chord member 28-28X.
Top right: hydraulic jack releasing erection cables.
Bottom right: erection of first lower chord members – Dawes Point.

Dr. J.J.C. Bradfield and W.H. Lush, University of Sydney Library's exhibition on the construction of the Sydney Harbour Bridge, date unknown.

Above: poster for the Festival of Britain, 1951. *Artwork by Abram Games, Festival of Britain, 1951.*

Below: photograph of the Skylon next to the domed exhibition building. *Photograph by Peter Benton, 1951. Creative Commons Attribution Share-alike license 2.0.*

The Skylon, 1951

"I saw a blade which rises in the sky held by hardly nothing at all."
Plaque commemorating The Skylon at the Festival of Britain, Jubilee Gardens.

Six years after the close of the Second World War, London was still in great need of redevelopment and investment. In an attempt to give Britons a sense of recovery, the Royal Society of Arts proposed a celebration to mark the centenary of the Great Exhibition of 1851. The Festival of Britain celebrated art and architecture through to science and industry, reaffirming Britain's resilience and acting, as put by the Festival Director, as a "*tonic to the nation*".

A lasting symbol of the Festival is The Skylon, a "*cigar-shaped pencil*", which rose to a height of nearly 295ft, its base held approximately 49ft from the ground. This towering structure consisted of a steel lattice, tapered at its ends, and was suspended by cables from three smaller steel beams. Each beam had a cable which ran to the Skylon's base and another which ran its top, acting as a stay. These cables were designed to be barely visible at a distance, making The Skylon appear to hover above the South Bank.

The high tension cables which held the iconic Skylon were tested for strength by David (Jnr) in the Kirkaldy Testing and Experimenting Works. Despite there being no direct record of the tests conducted, it is likely that lengths of the full size cables were tested in Kirkaldy's Universal Testing Machine. There still exist grips suitable for the testing of the cables but it remains unknown if they were actually used in the test. It is unlikely that Kirkaldy tested each of the individual wires within the cables, this most probably having been done by the manufacturer. However, the Works would have been an obvious choice for the testing of the full cables, given its proximity to The Skylon's site on the South Bank. There remains an irony that the Kirkaldy name, associated with quality and strength, is affiliated with a structure which "*had no visible means of support*".

The Skylon at the "Festival of Britain", 1951
British Official Photograph, Museum of London, 1951.

A Tonic for the Nation?

The Skylon was intended to elicit optimism in the public at a time when Britain was recovering from the Second World War. However, reactions to the short-lived icon were mixed. Some saw the fact that the Skylon had "no visible means of support" as an analogue of the British economy at the time. This swiftly become a well versed joke in the country. Others saw the monument as "a huge, lively joke, a tribute only to the spirit of nonsense and creative laughter".

For the most part, onlookers were impressed by the Skylon and experienced the positive message it was constructed to communicate. Among the most memorable sentiments are the following:

"This is the showpiece of the exhibition…By day or night it floats 300 ft. into the sky in isolated fantasy."

Colin St. C. Oakes in Builder

"Like the Eiffel Tower, it served no practical purpose except to make people stop and stare and wonder."

Barry Turner in Beacon for change : how the 1951 Festival of Britain shaped the modern age

De Havilland Comet Crash, 1954

By Tim Crichton

BOAC Flight 781

> *"It will perhaps be some consolation to the relatives of those who lost their lives to feel that good may come out of evil in this way"*
>
> Sir Lionel Heald QC, Attorney General

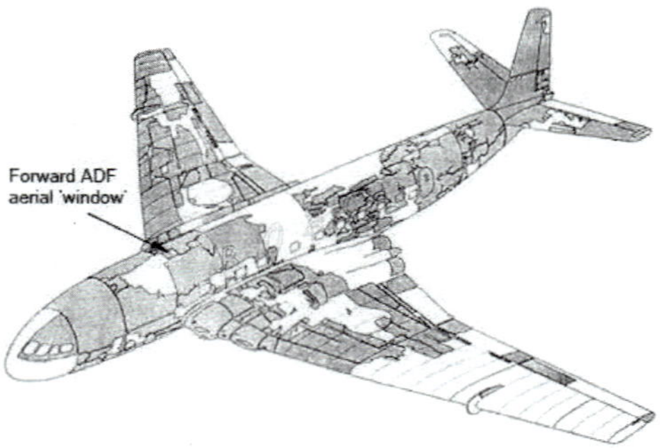

Far above: prototype de Havilland DH 106 Comet 1 G-ALVG in flight. BOAC's "Speedbird" symbol is displayed on its nose. *R.A. Scholefield, RAScholefield Collection, Airliners.net, 1949.*

Above: diagram showing the amount of wreckage recovered from the crash with forward aerial window indicated. *Unknown, Air Disasters by Stanley Stewart, Arrow Books, 1986.89. Creative Commons Attribution-Share Alike 3.0 Unported.*

Towards the end of the Second World War, aircraft survived longer before being shot down, however, air forces across the world began to experience losses of aircraft which could not be explained by enemy action or collision. These were early experiences of fatigue in airframes.

In 1949 the air registration board in England insisted that aircraft cabins withstand twice the pressure (2P) differential between the pressure at high altitude and the internal pressure for the passenger cabin. This was considered appropriate to ensure robustness of the cabin and also give a margin for fatigue. Further, airframes had to be tested to 1.33P with no permanent deformation.

The Comet aircraft was developed from a specification laid down by the UK Government's Barabazon Committee which was formed to determine Britain's need for commercial aircraft after the Second World War. The specification stipulated an airframe with a pressurised cabin capable of flying the Atlantic carrying 1 ton of payload at 400 miles per hour. Sir Geoffrey de Havilland, a member of the committee, influenced the specification to include that the aircraft should be turbojet powered. The de Havilland aircraft company had an enviable record in innovative and successful design and was awarded the contract to design and build the aircraft. Commercial considerations changed during the design process and it was decided that the aircraft should accommodate 24 passengers. The aircraft first flew on 27 July 1949.

The de Havilland Company was very rigorous in its design and testing of the new design. The fuselage was designed to 2.5P and tested to 2P. Working closely with the Royal Aircraft Establishment at Farnborough, Comet centre section and wings were subjected to loads in continual cycles throughout 1949-1951. Smaller components were tested in decompression chambers, and larger sections of fuselage were immersed in a specially-built water-tank. If a component ruptured in air, it would be blown to bits. Water acted as a damper, allowing the pressure to escape through the crack or tear, but keeping the rest of the structure intact. This made root cause analysis much easier. After rigorous testing and proving, the aircraft first went into commercial use on 2 May 1952 and initial operations of the aircraft were successful. Early incidents with aircraft were due to aerodynamic and structural issues.

On 10 January 1954 aircraft G – ALYP crashed into the sea near Elba. The airline immediately grounded its fleet of Comet airliners and a committee (the Abell Committee) was formed to investigate six possible aerodynamic and structural causes for the accident. This committee concluded that the most likely cause of the accident was fire. During the period of the Abell inquiry the Royal Navy had found the wreckage of the aircraft and recovered a large portion of the airframe. Commercial and political pressures caused the aircraft to return to service on 23 March 1954, however, the loss of G-ALYY on 8 April 1954 caused the whole fleet to be grounded and a large investigation board formed under the direction of the Royal Aircraft Establishment. On 19 October 1954 a committee was formed under the chairmanship of Right Honourable Lord Cohen and this tasked Sir Arnold Hall, Director of the RAE to perform more detailed investigations.

Above: a part of the Comet's fuselage being salvaged off the island of Elba.

Above: Sir Geoffrey de Havilland.
Unknown, San Diego Air and Space Museum Archive, date unknown

Left: wreckage reconstruction of G-ALYP.
Unknown, Ministry of Transport and Civil Aviation, 1954.

De Havilland DH-106 Comet 1 G-ALYP (closest), the first production Comet 1, with prototypes G-ALVG and G-ALZK (furthest) flying in the background.
R.A. Scholefield, RAScholefield Collection, Airliners.net, 1951.

De Havilland DH-106 Comet 1 G-ALYP at Heathrow North preparing for a scheduled flight.
R.A. Scholefield, RAScholefield Collection, Airliners.net, 1952.

De Havilland Aircraft Company

The de Havilland Aircraft Company is renowned for its innovations and shaping of the aircraft industry in the early to mid-20th century. It was established following the closure of Airco by the BSA (Birmingham Small Arms Company) in 1920, where Geoffrey de Havilland had worked for several years as technical director and chief designer. With his friend George Holt Thomas, de Havilland bought premises at Stag Lane Aerodrome in Edgware and established the company De Havilland Aircraft Company Ltd. Initially, the company focussed on small single and two-seat biplanes, using the Gipsy engine designed by de Havilland. These included the Gipsy Moth, Tiger Moth and prototype DH-60 Moth. The Moth series continued into the 1930s with the Cirrus and Hornet Moths, amongst others.

The de Havilland brand was also recognised for its design of high-performance aircraft, most notably the twin-piston-engine DH-88 Comet racer. The DH-88 gained this notoriety in 1934, when it achieved victory in the MacRobertson Air Race, a course between England and Australia.

During the Second World War, de Havilland's success with high-performance aircraft was of particular significance. Demand for powerful fighters resulted in the Mosquito, which continued to utilise the firm's wood construction techniques. Later, de Havilland created the Hornet fighter, making him one of the first-movers into metal construction methods. In this period, plans the Vampire were also drawn; an aircraft that would employ the recently conceived jet engine. The Vampire entered the RAF service in 1945 and was its first aircraft to be powered by a single jet engine.

The height of de Havilland's fame, however, came with the much anticipated DH-106 Comet, the world's first production commercial jetliner. Sadly, the failure of the Comet 1 in 1954 led to de Havilland being overtaken by his competition, notably the Boeing 707 jet. By the time the improved Comet 4 was released in 1958, its competitors were cheaper and faster. This led to the decline and eventual selling off of the firm.

Despite its collapse, the name de Havilland is still one recognised the world over for originality, experimentation and a certain amount of style.

Right: Geoffrey de Havilland sitting in the cockpit of his self designed and constructed aircraft number 2, 1911.
Unknown, Dera Farnborough , Imperial War Museum, 1911.

De Havilland D.H.91 Albatross, c. 1937.
Unknown, Charles Daniels Photo Collection album "British Aircraft", San Diego Air & Space Museum Archives, date unknown.

De Havilland DH-100 Vampire, 1951.
Unknown, Air Force Museum , 1951.

De Havilland DH-60 Moth biplane, c. 1929.
Unknown, John Oxley Library, c. 1929.

De Havilland DH-98 Mosquito, c. 1940.
Unknown, Canadian Forces, date unknown.

De Havilland DH-88 Comet, 1936.
Unknown, British official photographer, 1936.

De Havilland DH-95 Flamingo, c. 1938.
Unknown, Charles Daniels Photo Collection album "British Aircraft", San Diego Air & Space Museum Archives, date unknown.

Above: painting of Sir Geoffrey de Havilland.
Unknown, San Diego Air and Space Museum Archive, date unknown.

Below: prototype de Havilland DH 106 Comet 1 wearing its initial "B-Conditions" G-5-1 in flight (re-registered G-ALVG). The image was taken nine days after its maiden flight. G-ALVG was broken up at RAE Farnborough in July 1953.
R.A. Scholefield, RAScholefield Collection, Airliners.net, 1949.

Hall's team considered fatigue as a possible cause of the accident and commenced pressurisation tests on a Comet fuselage, G- ALYU in a water tank. After a total of 3057 pressurisation cycles (1221 flying cycles and 1836 cycles in the tank) the fuselage burst open and Geoffrey de Havilland and the aircraft's designer Bishop were called to Farnborough. The fuselage had ripped open at a corner of the forward left escape hatch cut-out. A further test reproduced the same results. At about this time the RAE had pieced together the parts of G-ALYP and found fatigue crack growth from the corner of a radio direction finding aperture in the metal fuselage.

Lord Cohen concluded that the Comet fuselage had been flawed, but that de Havilland had been working at the very limit of scientific knowledge. They had shown exemplary diligence, and were exonerated from blame.

Originally the various windows and apertures had been designed to be drill riveted and glued, however in the final construction process this method had been replaced by just punch riveting the holes. The punch riveting had caused minor cracks which, together with the square shape of the apertures, led to the growth of major fatigue cracks with disastrous consequences.

What was puzzling was why these flaws had not been discovered by de Havilland's own tests, or after the new ARB schedule in 1953. Sir Arnold Hall suggested that by applying 2P pressure initially, components had been strengthened, effectively "fatigue proofing" them. That is, by applying an extensive load and relaxing it, residual compressive stresses were introduced which masked any core weakness. The results were published openly in the de Havilland Gazette, and the company had discussions with British, American and other manufacturers and organisations, benefitting the entire industry, and in particular the next generation of aircraft. The Comet aircraft was redesigned with rounded windows and a thicker fuselage skin but the pioneering advantage of the aircraft had been overtaken by other passenger airlines.

"One of the most remarkable pieces of scientific detective work ever done"

Sir Lionel Heald QC, Attorney General

It should be noted that during the early stages of the investigations into the reasons for the failure of the airframe, Kirkaldy's Testing Works was tasked with carrying out a number of tests for the de Havilland Company. These tests revolved mainly around the previously experienced modes of failure of airframes where structural failure is caused by over stressing various parts of the airframe. These tests proved inconclusive although David Kirkaldy did voice concerns about the possibility of the window design leading to certain weaknesses.

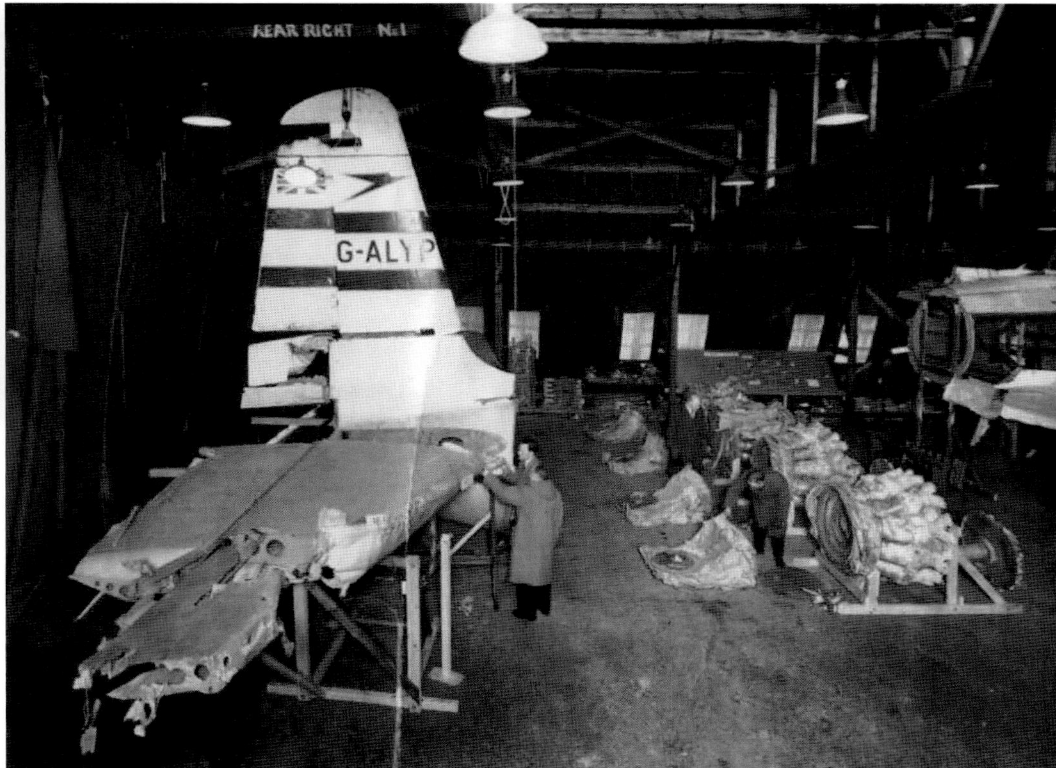

Left: G-ALYU in a water tank in order to simulate 3 hours of flight.
Unknown, Ministry of Transport and Civil Aviation, 1954.

Right: fragments of the crashed G-ALYP, c 1954.
© Science Museum / Science & Society Picture Library.

Stress Concentration

The de Havilland Comet crash of 1954 was a demonstration of the significance of stress concentration on the strength of a structure. It is often true that mechanical parts possess features which cause a drastic change in geometry (e.g. notches, holes, corners etc...). These areas of sudden change in topography lead to increases in localised stress fields, causing stress concentrations in the regions. Figure 1 demonstrates this effect on a stress field with the addition of a notch and a hole on a specimen, both of which are stress concentrators. These regions of stress concentration represent the locations where a structure is most likely to fail. In Figure 2, a photo of the failure within the de Havilland DH 106 Comet 1, one can see that the fracture occurred in the vicinity of a cornered window, acting as a stress concentrator.

Once more referring to Figure 1, it is clear that the effect of stress concentration is greater with the addition of a sharp corner as opposed to a rounded hole. A rounded edge, despite being a stress concentrator, represents a more gradual change in the geometry of a structure as compared to a sharp corner. The windows of modern aeroplanes are now rounded for this reason; in order to more evenly distribute stress around the structural feature.

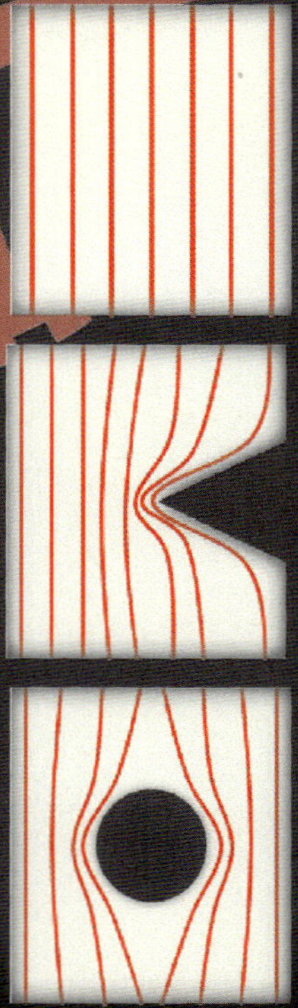

Figure 1 (above): diagrams showing the effects of notches and holes on stress distribution and concentration. The red lines represent the stress field. The closer together the lines, the greater the magnitude of the local field.

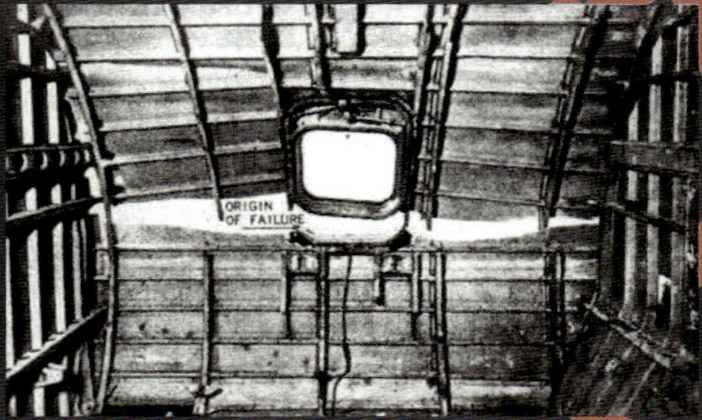

Figure 2 (right): photo of the failure at the forward escape hatch, at the port side of Comet.
Unknown, Ministry of Transport and Civil Aviation, 1954.

In order to quantify the stress concentration at points within a structure, a dimensionless factor (i.e. no units) called the Stress Concentration Factor (***K***) can be calculated. This factor indicates the extent to which stress is concentrated by finding the ratio between the peak stress (***σ_{max}***) within the structure and the reference stress (***σ_{ref}***) (i.e. the stress within the structure under the same loading without the stress concentrators):

$$K = \frac{\sigma_{max}}{\sigma_{ref}}$$

If there are no discontinuities within structure, the Stress Concentration Factor will be roughly equal to unity (≈1). However, if there exists significant stress raisers, the ratio with give a number much larger than unity (>>1).

A widely used method of determining stress concentration is photoelasticity. Photoelastic analysis is often employed as a more visual method, when a purely mathematical analysis becomes too complex. The effect of photoelasticity can be demonstrated by passing plane polarised light through a transparent material. When a scale model of a structure, made of such a material, is subject to stress (either internally applied by features within the structure or externally applied), a fringe pattern becomes visible (see Figure 4). This fringe pattern arises as a direct result of the photoelastic effect, which is a change in refractive index (***n***) caused by stresses. The refractive index refers to the ratio between the speed of light in a vacuum (***c***) and the speed of light within a material (***v***):

$$n = \frac{c}{v}$$

It is these changes of the refractive index within a scale model of a structure which creates the coloured fringe lines. As the relative refractive indices are proportionally linked to the relative stresses, the concentration of fringe lines can be seen as analogous to stress concentration in a specimen. The distortions in a stress field are, therefore, visually demonstrated on a structure.

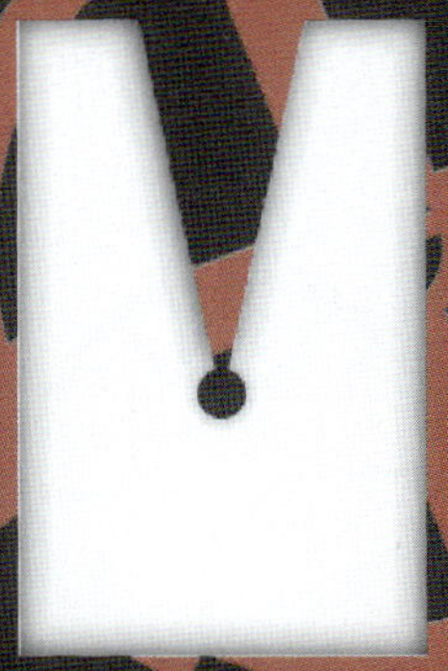

Figure 3 (above): diagram of a propagating crack in a structure, which can be halted by the addition of a round hole at its end. This hole, of relatively large diameter, reduces the stress concentration at the sharp end if the crack by the more gradual change in its geometry.

Figure 4 (below): a demonstration of photoelastic analysis on a protractor. The image clearly shows the concentration of fringe lines at the corners of the structure.
Photograph by Nevit Dilmen, 2004, Creative Commons Attribution-Share Alike 3.0 Unported license.

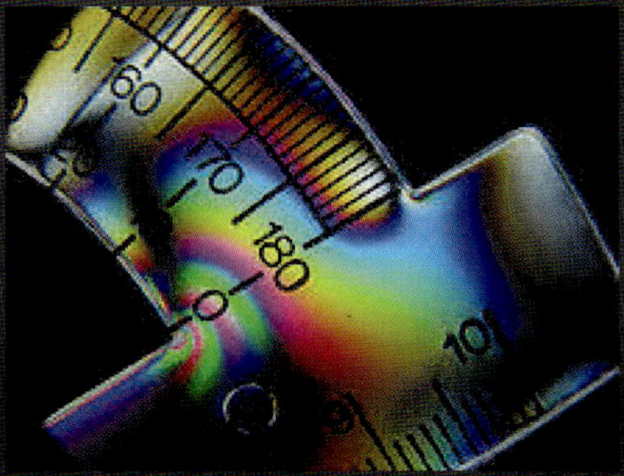

The Kirkaldy Testing Museum

The Formation of the Museum (1974 – present)

Above: Greater London Industrial Archaeology Society logo.
Courtesy of David Perrett, GLIAS.

Below: back of the Works on Prices Street in 1977.
Photograph courtesy of George Arthur, GLIAS.

Below right: original name plate on Kirkaldy's Machine in 1975. The plaque was removed and never replaced.
Photograph courtesy of George Arthur, GLIAS.

Following the close of Treharne and Davies in 1974, the Works ceased to function as an active testing works. In January of the subsequent year, members of the Goldsmiths College Industrial Archaeology class, headed by Dr. Denis Smith, and the Greater London Industrial Archaeological Society (GLIAS) made records of the machinery in situ and the building.

Prior to this visit, the Greater London Council (GLC) had recognised the architectural significance of the building but it was not until after the involvement of the archaeological societies in 1975 that both machine and building were listed as Grade 2 by English Heritage. This was the one of the first instances of a machine being listed as a part of a building.

"Above the doorway of the Kirkaldy Museum, carved in stone are the words "Facts Not Opinions". It makes me smile whenever I pass, because I live in a world so much in love with opinions, not facts..."
A.L. Kennedy, Facts not Opinions, Radio 4

In 1978, the first proposal to form a Museum was made, the team still being led by Denis Smith. The Kirkaldy Testing Museum was finally established in 1983 as a charity with generous guidance and funding from David Kirkaldy, the last of the Kirkaldys to operate the Works. From this point, a small group of trustees, initially headed by Denis Smith, has run the Museum alongside a handful of volunteers. The purpose of the Museum is to tell the little known story of the Kirkaldys within the context of the history of materials testing.

The importance of the Museum and the story it maintains was soon recognised. On 28th April 1987, Kirkaldy's Universal Testing Machine was honoured with the 10th Engineering Heritage Award of the Institution of Mechanical Engineers. The citation reads:

> *"This machine established the present-day system of materials testing and specifications of mechanical properties for engineering materials. "*

The building was sold by the Crown Estates in the early 1980s; the ground floor and the basement leased to the Kirkaldy Museum Trust. For many years the rest of the building was occupied by Waterman and Partners who were engineers and very supportive to the museum. They later left and the main lease is now in the hands of commercial landlords with whom the museum trust negotiated a new lease.

In 2015, the Kirkaldy Testing Museum and the Universal Testing Machine were upgraded to a Grade 2* by English Heritage.

> *"...the Kirkaldy Museum in London's Southwark Street, a museum of craftsmanship, safety and truth."*
>
> *A.L. Kennedy, Facts not Opinions, Radio 4*

Current Exhibitions

At present, the Kirkaldy Testing Museum hosts an eclectic range of testing machinery, dating from the mid-19th century to the modern day.

Above: 20th Century 60,000lb Riehle Machine on the ground floor of the Museum.

Below: Charpy Pendulum Tester (1916) in the basement of the Museum. It is believed to be the oldest functional Charpy Machine in existence.

Above: an Avery machine used by Irving Parachute Company (c. 1926).

Below and right: pianos being crushed in Kirkaldy's Machine.
Photographs courtesy of Afonso Guerra, Scrapclub, 2014.

Alongside many machines used by the incumbent Kirkaldys over 99 years, much of the collection has been donated. A large proportion of these donations have come from Imperial College, London. Kirkaldy's Universal Testing Machine continues to be the central attraction of the Museum and is used at each open day (traditionally, the first Sunday of the month) to test wrought iron bars in tension to submission. The bars used in this tensile test are cut from the original Victorian pavement gratings outside the Works, which were replaced by pavement lights in the 1990s. This demonstration of the Machine continues to provide visitors with a substantial jolt when the specimen finally breaks!

Further to the demonstration of Kirkaldy's Machine, there are a multitude of smaller machines still functional and frequently operated for visitors. Among these is an Avery machine (c. 1926) used by Irving Parachute Company to test the strength and extension of parachute lines in tension. Today, it is used to test plastic parcel strapping to submission and can be used by visitors. Also in use is a 1916 Charpy Pendulum Tester, thought to be the oldest of its kind in existence. This machine measures the energy absorbed when a pendulum impacts upon a specimen by the reduction in height of the pendulum when it swings back from its initial angle. The Charpy is one of two Pendulum Impact Testers within the Kirkaldy's collection and one of many readily operational machines in the Museum.

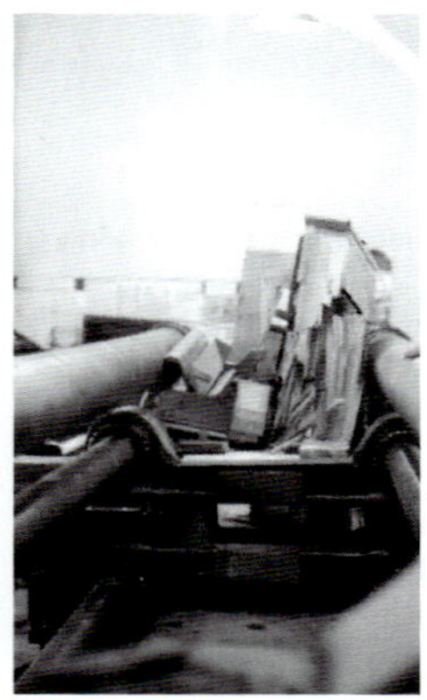

Collaborations and Projects

The unique environment and history of the Kirkaldy has attracted a multitude of quirky and exciting projects. In June 2014, the Museum collaborated with Scrapclub, a group dedicated to the artistic value of destruction. The group brought two second-hand pianos to the Museum to test to destruction in compression in Kirkaldy's Universal Testing Machine. Both pianos were filmed for high definition sound and video.

Despite applying about 70 tons, the Machine was unable to significantly damage the cast iron frames of either piano, but they still made for a good show!

Piano crushing, however, was not the first artistically motivated event to take place at the Museum. In September 2013, as part of the annual MERGE Festival, the groups Oscillatorial Binnage and Radio Resonance 104.4 FM collaborated in order to create the audio artwork "Music at Breaking Point". Both groups work with experimental acoustics and used the machines in the Museum as electro-acoustic musical instruments. The sound piece was entirely improvised in front of on looking visitors who watched the artists attach amplifiers to strained wood, vibrating metal grilles and stretched piano strings, amongst many other materials under stress. Oscillatorial Binnage recorded these audio clips to form the components of a mechanical synthesiser and created a musical composition as the conclusion to their work. Under such high amplification, the straining rattles and snaps of the tested materials did seem to possess an unworldly musical quality and varied widely in pitch and cadence. The highlight of this sonic exploration was the use of Kirkaldy's Machine to amplify the gradual creaking and, eventual, submission of a plank of wood. The piece was recorded for and broadcast by Radio Resonance, the world's first radio art station.

Above: a collection of images from "Music at Breaking Point".
Photograph courtesy of Tommophoto.com, 2013.

Left: one of James Capper's sculptures being stretched in Kirkaldy's Machine.
Photograph courtesy of Tommophoto.com, 2013.

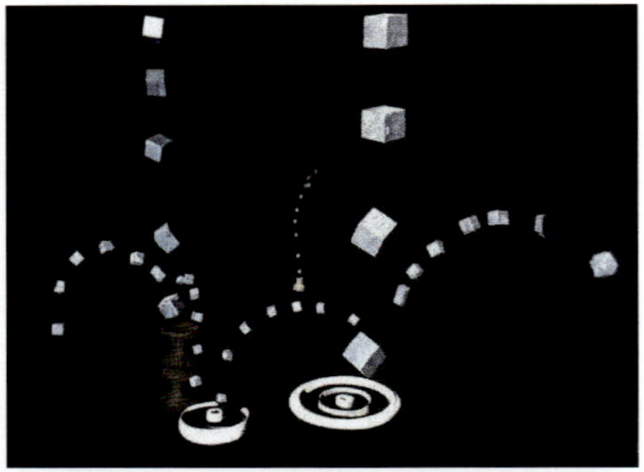

Above: still images from Trope's "Chain Reaction" animated artworks.
Photographs courtesy of Tommophoto.com, 2015.

Opposite: an photograph of the arches in the basement of the Museum. Their authenticity is often popular with installation artists.
Photograph courtesy of Richard Gooding, 2016.

In October 2013, again in conjunction with MERGE, the artist James Capper stretched his ten steel artworks in Kirkaldy's Machine. The Machine deformed and contorted the pieces as the tensile force was applied creating a moving artwork which evolved with time. Some of the forms were tested to submission, but many remained intact up to the Machine's maximum load. Capper allowed the Kirkaldy to display these tested pieces alongside the specimens collected over its years of operation.

Most recently, and once again in-line with MERGE, the installation art group Trope acted as "artists in residence" at the Kirkaldy as they developed their concept called "Chain Reaction" in the basement of the Works. The group, who see themselves as "designer of choreographed time", used an ingenious method of selective illumination of a sculpture to simulate movement. This unique system of animation is called the D-Scope®, which is a new approach to kinetic artwork. The basement was plunged into complete darkness to allow the dynamic lighting to be most effective. The hypnotic combination of artificial motion and audio effects truly immersed visitors. Among the artworks were swinging hooks, tumbling concrete cubes and shattering chains; giving elegance to crude and commonplace items at the Works.

The Museum has hosted a plethora of innovative and creative projects, all inspired by the captivating story of Kirkaldy and his Testing Works. In its 150th year, the Works remains a hub of intellectual, yet unorthodox, thinking and invention.

"An experience that takes you outside of reality…"

visitor to the Trope installation

The Kirkaldy Family Tree

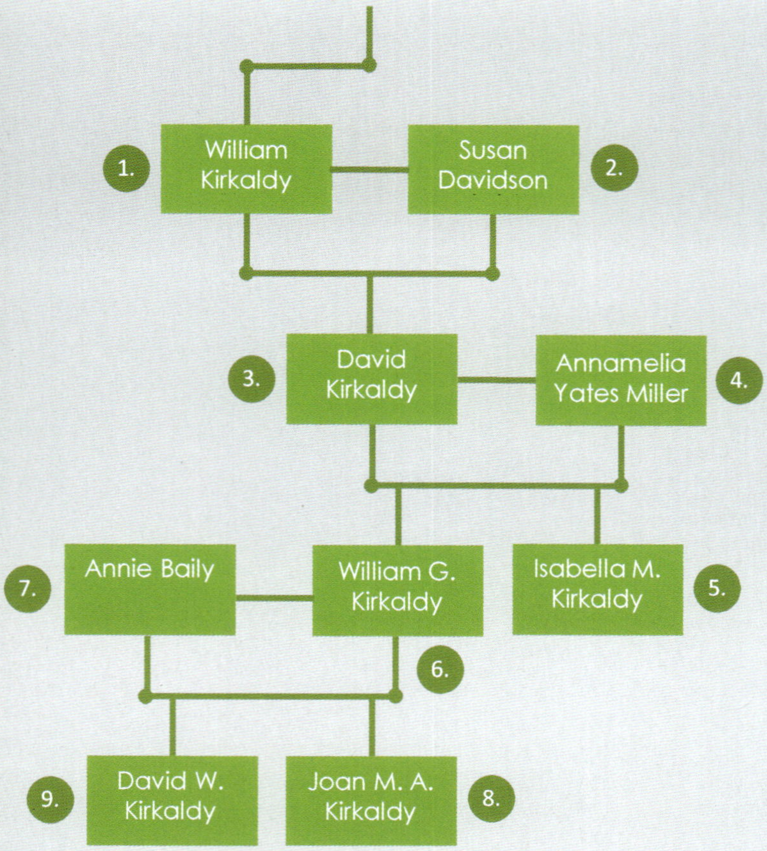

1. **William Kirkaldy (1786 - unknown):** the son of a John Kirkaldy, he was married to Susan Davidson (2.) on 20th September 1813 in Dundee. He appeared in the 1841 census where he described himself as a Merchant in Dundee. His son, David Kirkaldy (3.), was born in 1820 in Forfarshire (Angus).

2. **Susan Davidson (1797 – unknown):** baptised in 1797 in Brechin, Forfarshire (Angus). She was married to William Kirkaldy (1.) on 20th September 1813 in Dundee. She gave birth to a son, David Kirkaldy (3.), in 1820 in Forfarshire (Angus).

3. **David Kirkaldy (1820 – 1897):** baptised in 1820 in Dundee and declared head of the household at age 30, when he appeared in the 1851 UK Census on 30 March 1851 at Bucklemaker Wynd in Dundee. David Kirkaldy and Annamelia Yates Miller (4.) were married in 1858 and gave birth to their daughter Isabella Miller Kirkaldy (5.) was born in 1861 in Anderston, Glasgow. Their son, William George Kirkaldy (6.), was born in 1863 in Anderston. He died in February in 1897 at age of 77 in Islington, London.

4. **Annamelia Yates Miller (1836 - 1866):** born in 1836 in Oban, Argyllshire and described as a "Female Servant" in the 1841 census. She and David Kirkaldy (3.) were married in 1858 and gave birth to her first child, Isabella Miller Kirkaldy (5.), in 1861 in Anderston, Glasgow. Her son, William George Kirkaldy (6.), was born 1863 in Anderston. The family appeared in the 1861 UK Census on 7th April 1861 at the address Corunna Street in Barony, Anderston. She died in November 1866 at age of 30 in Pancras, London.

5. **Isabella Miller Kirkaldy (1861 - 1917):** born in 1861 in Anderston, Glasgow. She appeared in the 1891 census at the age of 30 and the daughter of the head of the household at Carleton Road in Islington, London. Later she appeared in the 1911 census where she describes herself as a woman of "Private Means" at North Villas in St Pancras, London. She died in May 1917 at age of 56 in Pancras, London.

6. **William George Kirkaldy (1863 - 1914):** born in 1863 in Anderston, Glasgow. He was married to Annie Bailey (7.) on 4th January 1894 in St. Albans Abbey, Hertfordshire. His daughter, Joan Margaret Anna Kirkaldy (8.), was born 28th October 1899 in Islington, London. He appeared in the 1901 census when he was age 38 and the head of the household at Carleton Road in Islington, London. He described his occupation as a "Testing and Experimenting Engineer". His son, David William Henry Kirkaldy (9.), was born on 14th August 1910 in Hampstead, London. In 1911, he was reported to live at Aberdare Gardens in Hampstead. He died in May 1914 at age of 51 in Hampstead.

7. **Annie Baily (1864 - 1938):** born in 1864 in London and lived at The Noke Farm in St Stephen, St Albans, Hertfordshire. She and William George Kirkaldy (6.) were married on 4th January 1894 in St. Albans, Abbey. Their daughter, Joan Margaret Anna Kirkaldy (8.), was born 28th October 1899 in Islington, London and their son, David William Henry Kirkaldy (9.), was born 14th August 1910 in Hampstead, London. She died around August 1938 at age of 74 in Lewisham, London.

8. **Joan Margaret Kirkaldy (1899 - 1980):** born on 28th October 1899 in Islington, London. She is recorded to have lived with her parents at both Carleton Road in Islington and Aberdare Gardens in Hampstead. She died around May 1980 at age of 80 in Brighton, Sussex.

9. **David William Kirkaldy (1910 - 1992):** born on 14th August 1910 in Hampstead, London. He is recorded to have lived with his parents at both Carleton Road in Islington and Aberdare Gardens in Hampstead. Later, he is reported to be a "Engineer Testing Of Material" in 1939 in Lewisham, London. In the 1939 UK Register, he was the head of the household at 12 Garlies Road in Lewisham, London. He died around November 1992 at age of 82 in Hove, Sussex.

Acknowledgements

This book owes much of its content to former chairman of the Kirkaldy Testing Museum, Denis Smith. His extensive research into the Kirkaldy family has proved invaluable, forming a basis and structure to current research investigations. Without his dedication to the formation of the Museum and its recognition as a national heritage site, the Kirkaldy family history would most likely have been lost in the annals of history.

~~~

We'd also like to offer our profound thanks to the Worshipful Company of Scientific Instrument Makers for their generous donation towards this publication.

~~~

A major supporter of the Kirkaldy Testing Museum in recent years has been bankside's annual MERGE Festival. Many of the up-and-coming artists and performers taking part in MERGE have chosen the Kirkaldy as the venue for their projects. Being associated with the festival has risen the profile of the Museum and made this book possible.

The artists who have taken part in festivals, such as MERGE, must also be credited for continuing to find inspiration in Kirkaldy's Testing Works. Their support has allowed Victorian engineering to have a place in the modern day and have made it accessible to a more diverse audience.

~~~

The commitment and enthusiasm of the volunteers who run the Museum and maintain Kirkaldy's visionary machine must also be acknowledged. Their support has allowed the Kirkaldy family history to be told for over 30 years and have secured its place in the future.

~~~

In truth, the Museum would not still be in operation without the vital support of our visitors. Our guests remain our primary source of funding in the upkeep and running of the Works. Their generosity and excitement about Kirkaldy's story has deeply touched all those who have operated the Museum over the last 30 years.

~~~

Finally, the team behind this book must receive thanks for the time and effort they have volunteered. It is this degree of dedication which will keep Kirkaldy's name alive for the next 150 years.
~~~

1st Floor, Kirkaldy's Testing and Experimenting Works, 1885

Museum of Fractures, 3rd Floor, Kirkaldy's Testing and Experimenting Works, 1885

Further Reading

Battle for the North by Charles McKean

Civil Aircraft Accident Report of the Court of Inquiry into the Accidents of Comet G-ALYP on 10 January 1954 and Comet G-ALYY on 8 April 1954

David Kirkaldy (1820 – 1897) and Engineering Materials Testing, by Denis Smith, The Newcomen Society, excerpt from volume 52, 1980 - 81

David Kirkaldy's Testing and Experimenting Works by Chris Rule

Failure Analysis: British Engine Technical Reports by F.R. Hutchings, and Paul M. Unterweiser

Illustration of David Kirkaldy's System of Mechanical Testing by William George Kirkaldy

Materials and Structures by E. H. Salmon

Report of the Court of Inquiry, and Report of Mr. Rothery, upon the Circumstances attending the Fall of a Portion of the Tay Bridge on the 28th December 1879.

Results of an Experimental Inquiry into the Tensile Strength and Other Properties of Various Kinds of Wrought Iron and Steel by David Kirkaldy

Substance of Civilization by Stephen L. Sass

Why Buildings Fall Down by Mario Salvadori

"[David Kirkaldy] was the terror of so many persons, that it is probable that at one time he was the best hated man in London, and he was the more hated that no joints could be found in his armour."

The Engineer, 1897

Bibliography

A. Emmerson, *Hydraulic Power in London*, Subterranea Britannica, 2015

A. Kludas, *Record breakers of the North Atlantic, Blue Riband Liners 1838-1953. London: Chatham, 1999*

C. R. Vernon Gibbs, *Passenger Liners of the Western Ocean: A Record of Atlantic Steam and Motor Passenger Vessels from 1838 to the Present Day. John De Graff, 1957*

D. Lyon, and R. Winfield, *The Sail and Steam Navy List: All the Ships of the Royal Navy 1815–1889, Chatham Publishing, London,* 2004

D. Perret, *An Uncertain Future for Kirkaldy's Testing Works in Southwark, Newcomen Links 226,* p**14**, 2013

D. Smith, *David Kirkaldy (1820 – 1897) and Engineering Materials Testing*, The Newcomen Society, excerpt from volume **52**, 1980 - 81

Donnachie, *Hydraulic Power Company (article is adapted from a talk given in November 1979 to the Lambeth and Southwark Archaeological Society)*, 2008

H.D. Young and R.A. Freedman, *University Physics*, 11th edition, 2003

History Channel, *'"Test Elephant" proves Eads Bridge is safe*

J. J. Colledge, B. Warlow, *Ships of the Royal Navy: The Complete Record of all Fighting Ships of the Royal Navy (Rev. ed.). London: Chatham Publishing*, 1969

National Park Service, *James B. Eads and His Amazing Bridge at St. Louis*, The Museum Gazette, 2001

R.W. Jackson, *Rails Across the Mississippi: A History of the St. Louis Bridge (Urbana: University of Illinois Press, 2001)*, 199

Sir Howard Roberts and Walter H. Godfrey (editors), *Southwark Street, Survey of London: volume 22: Bankside (the parishes of St. Saviour and Christchurch Southwark), chapter 15, page 90*, 1950

T. Turner, *Kirkaldy Testing Museum notes: 60 years after the war — nostalgia for the 1950s*, GLIAS, 2005

University of St. Andrews, *Mr James Valentine*

Unknown, *Archives In Brief 37 – A brief history of the Sydney Harbour Bridge*, The State Archives, NSW Government

Unknown, *Bridge Facts ,Pylon Lookout: Sydney Harbour Bridge*

Unknown, *Bridge History, Pylon Lookout: Sydney Harbour Bridge*

Unknown, *Minutes of the Proceedings of the Institution of Civil Engineers, Obituary William George Kirkaldy 1862-1914,* Volume **197** Issue **1914**, pp. **336-337** part **3**, 1914

Unknown, *Scottish Engineering Hall of Fame, engineeringhalloffame.org, 2016*

Unknown, *Sydney Harbour Bridge*, Australian Government, 2008

Unknown, *Sydney Harbour Bridge, Dorman, Long & Co. 1932. p.* **13**, 1932

Unknown, *The De Havilland Aircraft Company, RAF Museum website*, 2016

Unknown, *The Inst. C.E. Vol. CXXVIII, 1897*

Various authors, *Kirkaldy Testing Museum History*, 2016

Various authors, MERGE, 2016

Various authors, *Report of the Court of Inquiry, and Report of Mr. Rothery, upon the Circumstances attending the Fall of a Portion of the Tay Bridge on the 28th December 1879*, 1879

Various authors,, *Civil Aircraft Accident Report of the Court of Inquiry into the Accidents of Comet G-ALYP on 10 January 1954 and Comet G-ALYY on 8 April 1954*, 1954

W.G. Kirkaldy, *Illustration of David Kirkaldy's System of Mechanical Testing*,... 1891

Kirkaldy

Testing Museum

The Kirkaldy Testing Museum's logo

Index

G

H

J

K

L

M

N

O

P

R

S

T

U

V

W

The Authors

Far above: Lucy Hogarth, Tim Crichton and Colin Jenkins (from left to right).

Above: Hugh MacGillivray

Tim Crichton – Co Author

Tim works as an instructor for the RAF and has a background in engineering. He has acted as Treasurer and as a director at the Museum for over fifteen years. In addition to acting as Treasurer, he is a skilled tour guide of the Kirkaldy with a particularly keen knowledge of aircraft and their role in the Work's history.

Lucy Hogarth – Lead Author and Project Manager

Lucy is currently a Physics PhD student at UCL, studying for a four year course. She became involved with the Kirkaldy Testing Museum at the age of twenty and has acted as a volunteer and director for nearly five years. Her duties tend to focus on visitor experience and research. This book was originally conceived by Lucy in 2015.

Colin Jenkins – Co Author

Colin is employed in the field of computer software testing, having studied Physics at University and obtaining a PhD in the subject. He has now been involved with the Kirkaldy for about fifteen years and, at present, acts as Company Secretary as well as an enthusiastic guide and demonstrator at the Museum. Colin also maintains the Museum's website.

Hugh MacGillivray – Co Author

Hugh has been involved with the Kirkaldy Testing Museum for over thirty years, having only recently retired from his role as Chairman in 2016. He has worked as a researcher and lecturer of Materials Science at Imperial College London for much of his career and still conducts consultancy work in the field.

Cover photograph courtesy of Afonso Guerra, Scrapclub, 2014.

Artwork by Lucy Hogarth